NEW INSIGHTS INTO MORPHOMETRY STUDIES

Edited by **Pere M. Parés-Casanova**

New Insights into Morphometry Studies
http://dx.doi.org/10.5772/66563
Edited by Pere M. Parés-Casanova

Contributors

John Ndegwa Maina, Cleber Mansano, Beatrice I. Macente, Kifayat U. Khan, Thiago Matias T. Do Nascimento, Edney P. Da Silva, Nilva Kazue Sakomura, João Batista K. Fernandes, Tamira Elul, Anokh Sohal, James Ha, Manuel Zhu, Fayha Lakhani, Kavitha Thiagarajan, Lauren Olzewski, Raagav Monakrishnan, Sergey Kotov, Pere M. Pares-Casanova

Published by InTech
Janeza Trdine 9, 51000 Rijeka, Croatia

Publishing Process Manager Martina Blecic
Technical Editor SPi Global
Cover InTech Design team

First published July, 2017
Printed in Croatia
Legal deposit, Croatia: National and University Library in Zagreb

Additional hard copies can be obtained from orders@intechopen.com

New Insights into Morphometry Studies, Edited by Pere M. Parés-Casanova
p. cm.
Print ISBN 978-953-51-3365-0
Online ISBN 978-953-51-3366-7

For EU product safety concerns:
IN TECH d.o.o., Prolaz Marije Krucifikse Kozulić 3, 51000 Rijeka, Croatia,
info@intechopen.com or visit our website at intechopen.com.

Meet the editor

Pere M. Parés-Casanova graduated in Veterinary Medicine at the "Universitat Autònoma de Barcelona" (Catalonia) in 1987. He holds a Diploma in Public Health from the "Escuela Nacional de Sanidad" (Madrid, Spain), a Master's degree in Animal Production from the "Universitat Autònoma de Barcelona," and a PhD degree from the same University. From 2016, he is a lecturer and course coordinator of Gross Anatomy and Imaging Diagnostics at the Department of Animal Science in the School of Agrifood and Forestry Science and Engineering of the University of Lleida (Catalonia). His current research activity concerns the structural study of domestic mammals based on morphometrics techniques, with special emphasis to pure breeds. Evaluation of the morphological effects of extreme paedomorphy and asymmetries on "toy animals" has been recently his approach, too.

Contents

Preface

There has always been a need to understand the reality "through figures." Morphometry is a set of techniques, procedures, and computer resources that allow, thanks to a software of image analysis, to objectivate figure parameters of the object of study. A body figure is, in fact, a mathematically descriptive curve.

In its beginnings, morphology focused on simple descriptions of the observed structures. Initially, descriptions and comparisons were made qualitatively, i.e., compared with some easily recognizable form, using, for instance, terms such as "circled," "elongated," "fusiform," etc.

At the beginning of the twentieth century, the transition from descriptive studies to a more quantitative field occurred. Studies started to base on the analysis of the differences of their linear dimensions (measures, distances, angles, and proportions as variables). These measures were initially analyzed by univariate and bivariate statistical methods. The mathematical approximation to morphology originated directly from Francis Galton (1822-1911), Karl Pearson (1857-1936), Walter Frank Raphael Weldon (1860-1906), and Ronald Aylmer Fisher (1890-1962), to cite some. They developed methods of analysis with the aim of describing morphological variation patterns of intra- and intergroup. With technological advancement , their study acquired greater complexity and began to use multivariate statistical analysis—component analysis, canonical variables, discriminant function groups, etc. This approach is currently known in biometrics as "traditional morphometry."

Traditional morphometry provides the investigator of a set of analytical techniques to quantify morphological variation and to study the components that alter such variation. From variance-covariance matrices, it builds distances and uses multivariate analysis to all the original variables, each of which gives an account of a portion of the original variance. Conventional analyses are divided into those used for unique samples, without an assignment "a priori" of individuals in groups previously defined (principal components analysis), and those used for the analysis of two or more samples (discriminant function analysis).

The new variables derived from traditional morphometry correspond to major components or factors, with a number of useful properties:

1. They are orthogonal, that is, the values of each component are not correlated with the values of the other components.
2. The components are arranged in descending order according to the original variance percentage by which each responds, i.e., the first component will be the one which will see the highest percentage of the original variation.
3. The sum of the variance explained by each component will equal the sum of the variances of the original. These properties mean that summarizing the initial data in a few dimensions, but without losing the original information, so one can represent the better study objects and their relations on the basis of the characters studied.

Morphological analyses are currently widely used in different biological and non-biological fields, from microbiology to zoology and botany, paleontology and medicine, and even geology and hydrology.

This text is distinctive because of its emphasis on diversity issues. Each chapter shows brilliant applications in the study, analysis, and quantification of the variations in biological morphology. In the chapter "Morphometrics in Developmental Neurobiology: Quantitative Analysis of Growth Cone Motility In Vivo," there are fixed quantitative values for different morphological and motility parameters for growth cones on the optic tract of the African clawed frog (*Xenopus laevis*). In the chapter "Morphometric Growth Characteristics and Body Composition of Fish and Amphibians," authors describe that the relative growth and allometric coefficients of body components of fish and amphibians are described, classifying the growth of their corporal components. In "Application of Morphometric and Stereological Techniques on Analysis and Modeling of the Avian Lung," the researcher considers the versatility of stereological techniques in analyzing the avian lung. And, in the chapter "MRI Morphometry of the Brain and Neurological Diseases," the author studies brain structures in multiple sclerosis, Parkinson's disease, and cerebrovascular diseases. The characteristics relating to the nature of the primary data and the application of statistical treatment are absolutely different between all these studies, but all of them demonstrate the power capacity of modern morphometry to cope with a range of topics.

While reading of the book, it will appear clear to the reader that the technique is a powerful tool assessment objective which changes in form, presenting a wide research applicability. As an integrative discipline, morphology is well suited to show the workings of science itself.

This book, *New Insights into Morphometry Studies*, is written for you to increase your appreciation of morphometry applied on quite different fields and with different goals. With the lens of flexible techniques and mastery of concepts, all authors thread brilliantly a variety of material within a very different theme. Hence, the book can be viewed as an advanced text that should appeal to a wide audience, from researchers to medical practitioners who are searching for novel applications of morphometry, as well as to specialists in each considered chapter field. In conclusion, every advanced scholar interested in morphometrics will find this monography a must-read.

Pere M. Parés-Casanova, DVM
Lecturer
Dept. of Animal Science, University of Lleida
Catalonia, Spain

Section 1

Biological Sciences

Introductory Chapter - Morphometric Studies: Beyond Pure Anatomical Form Analysis

Pere M. Parés-Casanova

Additional information is available at the end of the chapter

http://dx.doi.org/10.5772/intechopen.69682

Morphometrics (or morphometry)[1] refers to the study of shape variation of organs and organisms and its covariation with other variables [1]: "*Defined as the fusion of geometry and biology, morphometrics deals with the study of form in two- or three-dimensional space*" [2]. *Shape* encompasses, together with *size*, the *form* in Needham's equation (1950) [3], two aspects with differing properties.

Scientific production in the morphometric field has increased dramatically over the last few decades. I do not doubt that largely this has resulted from easily available and (usually) fairly comprehensive computer programs, cheaper and more powerful personal computers, and more specialized and less expensive equipment for raw data acquisition: "*Fortunately, the morphometric community is replete with theorists who also generate software, and thus numerous packages are available*" [4].

Therefore, in addition to the "classical" tools for obtaining data (such as images), there is currently a wide spectrum of very advanced technology available, making measurements of any type easier, with more resolution, three-dimensional, less invasive and more complex: computed tomography, magnetic resonance imaging, ultrasound, surface scanners and other three-dimensional data-collection devices, scanners.[2] An example of this "new technological age" is the estimation of body surface area (BSA). The estimation of BSA can be traced back to 1793, when Abernathy directly measured the surface area of the head, hand, and foot in humans using triangular-shaped paper, estimating the remaining segments of the body using linear geometry [5]. Similarly in animals, initial BSA data were obtained by pasting strips of strong manila paper, gummed on one side, to the hair of the animals [6] or rolling a revolving metal cylinder of a known area, attached to a revolution counter [7]. Recently, however,

[1]From the Greek μορφή, *morphe*, meaning "form", and –μετρία, *metria*, meaning "measurement." The term "morphometrics" seems to have been coined in 1957 by Robert E. Blackith from Dublin University, who studied the subject in relation to locusts [1].
[2]No single type of imaging is always better; each has different potential advantages and disadvantages, and obviously their interpretation is subject to the hypothesis at hand.

complex techniques, such as computed tomography have been applied [8], and these have undoubtedly improved the quality (precision, ease) of data (and, frankly, I cannot imagine a live ferret being wrapped in a sheet of paper to estimate its BSA!).

A personal comment is in order here. These considerations have not been developed according to any deeper theoretical considerations. They are mainly based on personal experience of working with morphology in different contexts. Their aim is to provide an intuitive overview of how and for what purpose morphology can be applied, rather than attempting to formulate a strict thesis. Perhaps, needless to say, this is a text aimed at presenting certain personal ideas about morphometrics and morphology, not an attempt to give an exhaustive presentation of the literature on the topic. The bibliography presented is simply for things to make more sense and to demonstrate how I justify some assumptions on conceiving the ideas set forth.

Let us continue. Current software for morphometry can analyze data whatever their origin, and normally, it allows the construction of relevant images (the role of visual representations is very important in morphometrics, although algorithms sometimes cannot show completely accurate results, for instance, because they are not well adapted to a discrete framework).

Morphometrics was initially performed on organisms (*"Morphometrics is simply a quantitative way of addressing the shape comparisons that have always interested biologists"*) [9], extracting information by means of mathematical operations. Tools of morphometrical methods initially applied to study merely form (size + shape)[3] can be applied to other nonbiological fields. In this context, "morphometrical analysis" refers to the analysis of form within the particular scientific discipline where this term is used, including nonbiological forms. Many of the morphometrical concepts can, however, be generalized to encompass nonbiological hypotheses, and their applications are not currently restricted to biological uses. We now therefore have many branches of morphometrics which have emerged as a praxis of their own, such as "geomorphometry" [10] and "archaeometry" [3]. For a wider vision of morphology applications, it is recommended to read Zwicky's publications, which are listed on the website of The Fritz Zwicky Foundation (FZF) at: http://www.zwicky-stiftung.ch/index.php?p=6|8|8&url=/Links.htm. Furthermore, current morphological mathematical tools have similar advantages when applied to the study of "other-than-form" traits: color [11], pigmentation patterns, textures, etc. This is also the case when applied to meristic (countable) characters (for instance, fin rays in fish, cephalic foramina in skulls, etc.).

With this availability of many computational facilitations and so wide a spectrum of applications, current morphometric research cannot simply be applied to such a wide range of fields, but also requires the combination of many disciplines. All of these factors add up to a complex task, which should not be beyond our power as ordinary scientists. Morphometrics increasingly calls for an integrative research approach, in addition to a good understanding of the mathematical or logical basis of the approach considered.

In summary, we can give many answers based on any motivation of measurement, not only form, the *morphé*, on biological bodies. The important question in morphometrical analyses is frequently more related conceptually to how and what we measure than to how we should

[3]*Shape* contains the whole geometry (i.e., proportions) of objects, but it does not always take into account the overall complexity of the geometry of the specimens [3].

http://dx.doi.org/10.5772/intechopen.69682

proceed mathematically. For instance, same samples measured by means of geometric morphometrics or lineal morphometrics show totally different results, although statistical multivariate analyses are similar (comparing, for instance, [12, 13], it is clear how results can change according to a mere difference in how crude data were obtained (obviously I refer to technique, not quality)).

Morphology[4] "*refer*(s) *to the study of the structural relationships between different parts or aspects of the object of study*" [14]. It therefore includes aspects of outward appearance (shape, size, structure, color, pattern, i.e., external morphology or eidonomy), as well as the form and structure of the internal parts, like bones and organs, that is, internal morphology (or anatomy)[5]. Not only internal traits but also other external traits can therefore be mathematically analyzed with morphometric methods. We then have a huge cloud of research in a completely morphological—rather than merely morphometrical—field: biological or nonbiological specimens, on form or more structural traits, etc. For instance, in a study of mine of 322 eggs belonging to different Catalan hen breeds and varieties (data unpublished but available upon request from the author), the mere analysis of shape (using 3 classic descriptors "egg surface", "egg volume" [15], and "shape index" [16]) allowed 3.7% of correct identifications. When the analysis included fresh weight (which could be interpreted as size), they increased to 18.0%; and when the traits studied included color (cream or tinted, white or brown), successful classification reached 20.8%. This is just an example of how results can be obtained by means of a production process—in some cases, a complex one—but which will be influenced by decisions on the hypothesis taken rather than by the mathematic algorithms concerned.

In conclusion, morphometrics, being a branch of statistics, must be viewed as a branch of morphology in the widest sense.[6] Also, on emphasizing the broad component of morphology, we do not rule out the significance of its mathematical component.

Author details

Pere M. Parés-Casanova

Address all correspondence to: peremiquelp@ca.udl.cat

Department of Animal Science, University of Lleida, Catalonia, Spain

References

[1] Reyment RA. Morphometrics: An historical essay, In: Elewa AMT, editor. Morphometrics for Nonmorphometricians, Lecture Notes in Earth. Vol. 124. Springer-Verlag Berlin Heidelberg

[4]From the Ancient Greek *don, morphé*, meaning "form," and λόγος, *lógos*, meaning "word, study, research".
[5]From the Ancient Greek ἀνατομή, *anatomē*, meaning "dissection", and –τέμνω, *témnō*, meaning "I cut".
[6]And with morphometric technique being dependent on images, would it be better defined as "morphography"? I leave it to the readers' consideration.

[2] Richtsmeier JT, DeLeon VB, Lele SR. The promise of geometric morphometrics. American Journal of Physical Anthropology. 2002;**45**:63-91

[3] Borel A, Cornette R, Baylac M. Stone tool forms and functions: A morphometric analysis of modern humans Stone tools from song terus cave (Java, Indonesia). Archaeometry. 2017;**59**(3):455-471

[4] Adams DC, Rohlf FJ, Slice DE. A field comes of age: Geometric morphometrics in the 21st century. Hystrix. 2013;**24**(1):7-14

[5] Daniell N, Olds T, Tomkinson G. Technical note: Criterion validity of whole body surface area equations: A comparison using 3D laser scanning. American Journal of Physical Anthropology. 2012;**148**(1):148-155

[6] Hogan AG, Skouby CI. Determination of the surface area of cattle and swine. Journal of Agricultural Research. 1923;**25**(419):419-432

[7] Elting EC. A formula for estimating surface area of dairy cattle. Journal of Agricultural Research. 1926;**33**(3):269-280

[8] Jones KL, Abbigail Granger L, Kearney MT, da Cunha AF, Cutler DC, Shapiro ME, Tully TN, Shiomitsu K. Evaluation of a ferret-specific formula for determining body surface area to improve chemotherapeutic dosing. American Journal of Veterinary Research. 2015;**76**(2):142-148

[9] Zelditch ML, Swiderski DL, Sheets HD. Geometric Morphometrics for Biologists: A Primer. Boston, MA : Elsevier Academic Press; 2004

[10] Guth PL. Drainage basin morphometry: A global snapshot from the shuttle radar topography mission. Hydrology and Earth System Sciences. 2011;**15**(7):2091-2099

[11] Hall-Spencer JM, Moore PG, Sneddon LU. Observations and possible function of the striking anterior coloration of Galathea intermedia (Crustacea: Decapoda: Anomura). Journal of the Marine Biological Association UK. 1999;**79**:371-372

[12] Parés-Casanova PM, Morros C. Molar asymmetry shows a chewing-side preference in horses. Journal of Zoological and Bioscience Research. 2014;**1**(1):14-18

[13] Parés-Casanova PM, Reig E. Directional and fluctuating asymmetries in Cavall Pirinenc Català breed molars. Journal of Animal Ethnology. 2015;**1**:10-18

[14] Álvarez A. Ritchey T. Applications of General Morphological Analysis. Acta Morphologica Generalis. 2015;**4**(1):1-40

[15] Havlíček M, Nedomová Š, Simeonovová J, Severa L, Křivánek I. On the evaluation of chicken egg shape variability. Acta Universitatis Agriculturae et Silviculturae Mendelianae Brunensis. 2008;**56**(5):69-74

[16] Altuntas E, Sekeroglu A. Effect of egg shape index on mechanical properties of chicken eggs. Journal of Food Engineering. 2008;**85**(4):606-612

Chapter 2

Morphometric Growth Characteristics and Body Composition of Fish and Amphibians

Cleber Fernando M. Mansano,
Beatrice Ingrid Macente, Kifayat Ullah Khan,
Thiago Matias T. do Nascimento, Edney P. da Silva,
Nilva Kazue Sakomura and
João Batista K. Fernandes

Additional information is available at the end of the chapter

http://dx.doi.org/10.5772/intechopen.69061

Abstract

Describing animal growth through the nonlinear models allows a detailed evaluation of their behavior, besides revealing important information of the response to a particular treatment. In this chapter, the parameters of mathematical models (Gompertz, Von Bertalanffy, Logistic and Brody) for live weight, feed and protein intakes, total and standard lengths and nutrient deposition are described systematically and comprehensively. Also the relative growth and allometric coefficients of body components in relation to body weight of fish and amphibians are described, explaining better the use of the allometric equation and classifying the growth of the body components.

Keywords: mathematical models, allometry, body components

1. Introduction

The growth of an animal is directly related to its weight gain, constituted by water retention, protein, fat, and minerals, the quantity of which may vary from organism to organism. The order of formation of tissues and bone, muscle or fat, depending on the physiological maturity, that is, the development of each tissue occurs in an isometric way, besides that each component stimulates its growth in different phases of the animal life [1, 2]. Through this sequence, the final target of the nutrients in the animal's body has been observed. It is therefore important to know the weight and/or age at which the body growth rate declines and most part of the nutrients goes to the adipose tissue due to the increased demand for energy expenditure [3]. Research is needed to determine the body growth in order to describe well the increased animal production. However, a proper animal's physical growth requires good maintenance conditions which are provided by the bone structure, whose development should be closely linked to the development of muscles for obtaining optimal body growth [4].

In aquatic animals, the adipose tissue may occur as individual deposits, like visceral fat existing in the form of body fat [5]; or less diffusely distributed in muscles, liver, skin, kidneys, lungs, bones, and connective tissues [6]. The fat deposits in tadpoles are acquired from the genes that are transferred to them from their parents [7]. However, other factors also contribute to the accumulation of fat in animals, such as diet and environmental conditions [8].

Animal growth is directly related to the feed it receives and the climate conditions of the region where it is found. Moreover, it is also associated with other factors such as the genetics, biotype, race, weight, age, and body state [9]. Emmans [10] stated that an ideal method of calculating nutritional requirements and indicating an animal's feed intake during its development is first to start discovering its growing potential. The nutrient requirements and development of an animal are fully interconnected. Through the understanding of these interactions, it is therefore, possible to find out the nutritional deficiencies, and to obtain the maximum animal performance while discerning the limits of production and making the appropriate changes to improve the animal productivity [11].

2. Animal growth characteristics

Growth involves an increase in the size of the animal, accompanied by the changes in body components, the latter being known as "development." The body components change in response to the age-related changes in cellular structures and functions [12]. Growth begins after fertilization of the ovum and ends when the body gains the adult weight [13].

Muscle growth in fish differs from that in mammals and continues for much of the life cycle [14] or fishes do not stop growing even after breeding. In mammals, Gómez et al. [15] described that weight gain is produced by three processes—the hyperplasia, an increase in the number of muscle cells; the hypertrophy, an enlargement of the cells; and metaplasm, the transformation of cells. Thus, animal growth is a cellular response to different internal and external factors.

In fish, three phases of muscle formation are distinguished: the first phase leads to the formation of embryonic muscle fibers that are grouped as undifferentiated myoblasts, which are the source of subsequent growth. In the second phase, the yolk sac larvae are observed, the differentiation of the germinal and proliferative zone of the myoblasts, where the dorsal and ventral apex of the myotomes is observed. Thirdly, the myoblasts on the surface of the embryonic muscle fibers are activated in a process that can continue for the entire life [15].

In amphibians, more precisely in bullfrog tadpoles, the life cycle is divided into three phases—embryonic, larval, and metamorphosis. The embryonic stage constitutes the period of fertilization and development within the egg. The second stage, larval, begins with the hatching of the egg and the entire period of development of the tadpole. In the third and last phase, metamorphosis, the tadpole changes into an adult amphibian [16]. Gosner [17], while taking into consideration the morphological changes that occur in the three phases, subdivided them into 46 developmental stages. The first embryonic phase includes the 1–25 stages; the second phase, of body growth and early development of the hind limbs, includes the 26–35 stages. During the stages 36–41, the stabilization of the corporal growth and the development of hind limbs occur and, in 42–46 stages, the metamorphosis ends up with the externalization of the forelimbs, reabsorption of the tail, and modification of the mandible. However, the bullfrog, like all anuran amphibians, is carnivorous during the adult (terrestrial) phase, generally requiring higher levels of protein in its diet than those of other eating habits [18–21].

2.1. Factors regulating the animal growth

The growth of an animal depends on the genotype-environment interaction and factors such as quality and quantity of food, management, and health status. Body weight and length are the main parameters for producers (breeders) to determine whether the feeding level is adequate or not. For the diet to meet rapid growth is essential to understand the relationships between the weight or length and growth, so if the food is insufficient for maintenance and growth, the latter may inhibit or cease altogether [22].

In addition to weight and feed consumption, growth is influenced by other factors, which often interact with the amount of feed and body weight. According to Hepher [22], these factors can be internal and external (environmental) factors. Thus, for example, some species show a clear difference according to sex ranging from 5 to 10% [23]. Dutta [24] describes that males of *Xiphophorus* and *Poecilia* reach a "specific size," however the females continue to grow after maturity while the rate of growth decreases over time. Another example is tilapia, where males grow faster than females, even yet in common carp (*Cyprinus carpio*) and eel (*Anguilla anguilla*), there is a greater growth of the female in relation to the male [22].

When the female has a lower weight than its male counterpart, the values of the initial growth rate, the inflection point and the asymptotic weight are smaller; nevertheless, it has less time to reach maturity. These differences of precocity between the sexes can be observed in the growth of the different tissues [23, 25], in some genetic characteristics, and in the physiological state of the animal. The growth of some fish decreases when they reach to sexual maturity. Some species of the genus *Oncorhynchus* and *Anguilla* migrate to spawn and die. On the other hand, the *Salmo salar* can repeat this procedure many times, that is, its individuals feed and grow between each spawn. In case of some tropical fish such as Heterandria, the growth is interrupted when the animal reaches a "specific size" [24]. External factors included that affect growth are the temperature, light, and water quality, which may interact with the genotype of the fish and amphibians and can induce variations in the muscle growth rate [26].

In frog culture, the time of production of the "imago," to reach slaughter weight can range from 77 [27] to 166 days [28]. The main interference factor is the temperature because it directly influences the metabolism of the animal. Similar to all anuran amphibians, the bullfrog is dependent on the temperature of the environment in which it is found [29]. Sometimes, an increase in the water temperature above the level considered to be optimal for bullfrog may influence growth performance. Braga and Lima [30], observed a better growth and weight gain of bullfrogs with a live weight between 37 and 90 g at the temperature between 25.1 and 30.4°C. Figueiredo et al. [31] already have observed the better performance parameters in bullfrogs weighing more than 100 g at temperature between 27.6 and 28.2°C. The environmental temperature also affected the adipose and hepatic tissue weights, presenting the higher values at temperatures of 27.27 and 26.81°C, respectively [32].

2.2. Mathematical models for describing animal growth

According to Tedeschi [33], models are mathematical representations of the mechanisms governing natural phenomena that may not be fully recognized, controlled, or understood. A mathematical model is an equation or set of equations which represent the behavior of a system, where there is a correspondence between the variables of the model and the quantities observed [34]. According to Dumas et al. [26], mathematical models are analytical solutions for the differential equations that can be adjusted to the growth data using nonlinear regression. Likewise, regression analysis uses the relationship between two or more quantitative variables, so that one variable can be assumed as a function of another. The main objectives of regression analysis are based on three purposes: description, control, and prediction [35].

The modeling process includes the definition of objectives, construction of a diagram to identify the main factor involved in the system to be modeled, formulate the appropriate mathematical functions, collection of the data to estimate the parameters, solving equations, evaluation and verification of the model and programming the simulation [36].

Growth in animals can be explained by mathematical functions. These functions can predict the development of live weight, which helps to evaluate the productivity of a breed under a specific breeding condition [15, 37]. Growth can usually be described and predicted using conventional mathematical models, since it does not occur in a chaotic way [26]. In order to understand the random variation between the measurements of an animal, growth curves can be used with the aim of adjusting and standardizing the variation of weight and age during the life of an individual.

Growth models have been used to provide a mathematical summary of the development of animal growth or its parts as a function of time [34]. The growth model expression is used to describe an analytical function described by a single equation: $y = f(t)$, where "y" is the response variable (weight) that depends on the functional relationship, which is established as a function of the independent variable "t" (time).

According to Thornley and France [34], growth models can be categorized according to the functional behavior "f" as curves describing a decreasing yield (Monomolecular), those which have a sigmoidal behavior with a inflection point (e.g., Logistic, Gompertz, and Schumacher) and those with a flexible inflection point (as Von Bertalanffy, Richards, Lopez, and Weibull).

Growth curves which involve a series of measurements of some interest over time (body weight, body composition, diameter, and longitude) [38] are usually adjusted under controlled conditions, and are the first steps in predicting nutrient requirements for animals of different genotypes [26, 39]. Moreover, they evaluate various parameters such as growth rate, maturity rate at different ages and weight at slaughter time and thus allow and help in establishing zoo technical breeding programs [15].

According to Brown and Rothery [40], each model has the ability to calculate an estimate of mean weight at maturity and early maturity periods. The closest asymptote is the weight at maturity, as a constant condition relative to a model for body composition under productive environments. Dumas et al. [26] showed that the growth trajectory of the animals presents an initial phase of acceleration, and the levels when the animal is close to its adult stage or induces its reproductive growth, being called the growth inhibition phase (**Figure 1**). Many species of fish, molluscs, crustaceans, and amphibians can even grow after reaching the maturity size and the final stage of growth presents a greater plasticity [26].

2.3. Models applied for growth assessment

To describe the growth in fish and amphibians, it is common to use nonlinear mathematical models. The most used models are included, the Brody, Gompertz, Logistic, Richards, and von Bertalanffy [40–44], however, there is a much larger range of functions that can be used to help in the simulation of body growth and body components such as, the scales, skin, viscera, fillet or nutrients such as, the protein, fat, and ash content. These functions are used in simulation models to estimate the body composition of animals at any stage of development, requiring little information on their growth and initial body composition [26].

These models contain several common parameters though there are variations regarding their interpretation and content, and are possible to associate any biological meaning to each of them [25]. Gompertz, $Y = A\exp(-\exp(-b(t - T)))$; Von Bertalanffy, $Y = A(1 - K \exp(-Bt))^3$; Logistic, $Y = A(1 + K\exp(-Bt))^{-1}$; and Brody, $Y = A(1 - K\exp(-Bt))$. The parameters used in these models are defined as: Y = measurement values (g or cm); t = experimental days; A = body

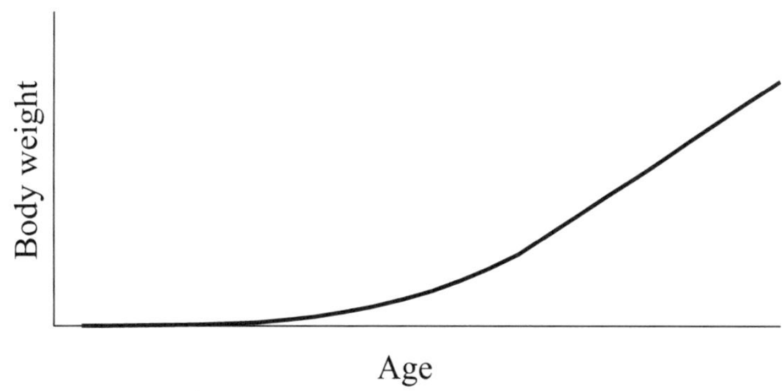

Figure 1. Typical growth trajectory in fish (*source*: Dumas et al. [26]).

weight or length at maturity; K = scale parameter with no biological interpretation for the Von Bertalanffy, Logistic, and Brody models; b and B = growth rate at maturity; T = growth rate at maturity for the Gompertz model, where it represents the day of maximum growth. These parameters can be estimated by the modified Gauss-Newton method, through the program of SAS by procedure "PROC NLIN" (nonlinear regression).

2.4. Assessment of the accuracy of mathematical models

According to Tedeschi [33], the evaluation of the accuracy of a model is an essential step in the modeling process which indicates the level of precision in the prediction adjustments. The evaluation of the model can and should proceed up to the level of the predicted results (upper level) and up to the level of the assumptions (lower level), while the parameters should be determined by the researchers. Unfortunately, this is not always possible, and some "tuning" or "calibration" of the parameters is usually necessary. A higher evaluation may consider model properties, such as simplicity, fit plasticity, applicability, and quality and quantity of prediction adjustment.

According to Santos et al. [45], to choose the model that best fits the data, the following criteria are considered: mean square of residue (MSR), coefficient of determination (R^2), and biological interpretability of parameters.

Some criteria can be used to select the models and describe correctly that which one is better for a given data. The most commonly used adjustment quality evaluators are determination coefficient (R^2) [46, 47], adjusted coefficient of determination (R^2aj.) [48], the mean squared error (MSE) [46, 47]; value of the Akaike criterion (AIC) [47, 48], value of the Bayesian information criterion (BIC) [47, 48]; convergence percentage (C%) [47, 48], the number of iterations (NI) [44, 45, 46]; mean absolute deviation of residues (MADR) [49–51], dispersion of the waste estimated by the models and the distribution of studentized waste [50].

The adopted set of adjustment evaluators should be fitting to assist in the decision making of the choice of the better model studied. Evaluation criteria for selecting an appropriate model should be well adopted, since information provided by fit quality assessors can indicate that which model is most appropriate to describe the body growth of a population [52, 53].

2.5. Mathematical models for evaluating animal growth

Each animal species has a particular growth curve where it should be in a suitable and non-limiting environment. The fact that several aspects such as maturity, composition, and deposition rates of body nutrients can interfere with the growth curve should be emphasized. Therefore, care must be taken in choosing the best model, since there are a lot of models that will fit one's data. However, attention should be paid about those who describe the growth of animals with greater precision and clarity according to their age. Once the wrong model has been chosen, the error will be reflected in future researches and feeding programs.

In the ongoing nutrition research, several studies on the application of mathematical models are available about amphibians and fish. Thus, some of the previous literature has been chosen to characterize their application in research studies. Next will be described the weight or length of the animal at maturity (A) and growth rate relative to maturity (K) of some species.

As for the parameter *A* that refers to the asymptotic weight, Amancio et al. [54] evaluated fitting of five mathematical models (Gompertz, Logistic, Linear Hyperbolic, Quadratic, and Logarithmic Quadratic) to describe the growth curve of genetically improved farmed tilapia (GIFT) Nile tilapia (*Oreochromis niloticus*). Nile tilapia fingerlings of the initial weight 2.4 g stocked in 20 concrete tanks of 2 m^3 with a density of 25 m^{-3} fish for a period of 180 days. An asymptotic weight of 763.6 g in the Gompertz model and 509.8 g in the Logistic model was reported. These values are lower than those found by Carvalho [55], who worked with several families of tilapia in a genetic improvement program, using the Gompertz model and found values of 3921.4 g (family 40), 4554.7 g (family 6), and 4613.5 g (family 53).

Similarly, the values for the *T* parameter were obtained, which refers to the age at the inflection point of 186.6 days of the Gompertz model and 208.2 days of the logistic model. These values also are less than those found by Carvalho [55], such as 495.4 days (family 53), 479.8 days (family 6), and 415.2 days (family 40). It is then realized that such differences of the variety between the Nile tilapia families imply that studies can still be carried out for the improvement of the variety in question, as well as the validation and concluding the fish growth, or even studies for the formation of a new variety.

In conclusion, Carvalho [55] concludes that Gompertz model was the one that better adjusted the characteristics evaluated in the study of growth curves of Nile tilapia. Similar result was seen by Hernandez-Llamas and Ratkowsky [56] and Katsanevakis and Maravelias [57] evaluating mathematical models to describe the fish growth. However, in a study carried out by Aguilar [58], using the Chitralada variety of Nile tilapia, a better adjustment of body growth rate for the von Bertalanffy model, Gompertz, and Logistic has been found. The other models also presented a satisfactory fit with the estimated asymptotic weight between 614.13 and 820.44 g. On the other hand, Costa et al. [59] used the Brody, von Bertalanffy, Logistic, Gompertz, and exponential models evaluating the growth of the Chitralada, GIFT, and red Nile tilapia lines and observed that the fit of the exponential model was the most adequate. In the present experiment, the Gompertz model was the one that better adjusted the data.

In the same way as evaluated for fish, in amphibians more specifically in bullfrog tadpoles, different models (Gompertz, Brody, von Bertalanffy, and Logistic) were applied for evaluation and simulation of their growth [50]. The values of the parameters found for each growth model adopted in weight and total length have been found in the study of Mansano et al. [50]. The Gompertz, von Bertalanffy, and Logistic, were the only featured models in which the convergence criterion was achieved; however, the Brody model did not converge for the observed data set of weight and length. A possible explanation may be that the model does not have an analytical solution of the normal equations, being the estimations of the parameters of the nonlinear models obtained by iterative algorithms [60]. The weight at maturity or asymptotic weight (*A*) found for the logistic model (8.90 g) was the one with the lowest value, followed by the Gompertz model (10.66 g), which was lower than that found by the model von Bertalanffy (13.36 g) [50]. For the total length at maturity, parameter *A* presented the same behavior. The simulated values for parameter *A* adopted in the study are biologically interpretable for bullfrog tadpoles.

The parameter of *B* of the Gompertz, von Bertalanffy, and Logistic models represents the growth rate relative to maturity, and the lowest value of this parameter represents the highest weight and total length at maturity [61]. In the study of Mansano et al. [50], the logistic model presented the highest value, among the three models that were presenting the lowest weight and total length at maturity. Inversely, the von Bertalanffy model presented the lowest *B* value for the parameter *B* and, consequently, higher weight and total length at maturity. In the same study of the evaluation of these models, the R^2 values found for all the models were excellent >0.98, with small differences for both live weight and total length. However, R^2 is not a good differentiator for choosing nonlinear models [60]. From the mean squared error (MSE) found for the models, it can be seen that there was no difference between the studied models, for both live weight and total length. In the absolute average deviation (AAD) evaluation, it was possible to verify that the von Bertalanffy and Logistic models underestimated the values. They presented lower values than those observed in the initial weight studies which are a serious error to be considered. Since for animals such as bullfrog tadpoles that have an initial weight around 0.1 g, it cannot be considered that an animal has a negative weight because it is biologically impossible.

In a study of captive bullfrog during its terrestrial phase, Pereira et al. [62] tested two nonlinear models. The authors have found quite different values among which the estimated value of *A* for live weight of 1051.5 g of the Gompertz model was considered high to represent the study period. Bullfrog specimens may reach this value throughout their life with more than 2 years. However, the estimated value for mean weight of 343.7 g of the logistic model was considered adequate for the growing period of the "Froglets" until the slaughter weight, since the frogs had an average weight of 214.56 g with 126 days. The adjusted K value for live weight of 0.0088 (g/day) by Gompertz model presented the same incoherence for the A value of the same model. Since it is believed that the bullfrog presents a maximum growth rate during the period of the 126 experimental days, estimed as 0.0313 (g/day) by logistic model with its maximum peak at 109th day of the experiment.

According to Pereira et al. [62], the logistic model presented a characteristic of estimating baseline values lower than the Gompertz model, underestimating the initial live weight in the mean of 4.12 g. This behavior was also observed in bullfrogs created in mini bays, where the logistic model underestimated the initial weight by 21.8 g [51], and the study performed in 294 days with frogs beyond the slaughter weight range. These values underestimated by any type of model provided can be considered that this value is not negative, since no animal is born with a negative weight. It is important to point out that the results found in the previous literature aimed finding the equations representing growth may vary among the various species of amphibians and the conditions adopted [6]. The choice of a suitable growth model is important, since it can have a decisive effect on the results of simulation of an ecological dynamics model. For example, the logistic model has been indicated to describe growth over short periods of time (days and months) and in environments that have some control such as nutrition [63].

The use of nonlinear models may have a wide application area, using the Gompertz model described by Mansano et al. [50], to describe the growth curve and body composition (protein crude, fat, water, and ash content) of bullfrog tadpoles [64] (**Table 1**). In addition to the evaluation and simulation of growth using Gompertz model, it was possible to verify that which of

http://dx.doi.org/10.5772/intechopen.69061

Variable	Diet	Parameter		
		Pm	*b* (per day)	t^*
Live weight (g)	ED	10.66 ± 1.0517[a]	0.0558 ± 0.0088	38.195 ± 2.2956
	CD	9.54 ± 0.4174[b]	0.0590 ± 0.0044	37.571 ± 0.9918
P value		0.0028	0.3628	0.3020
Total length (mm)	ED	120.0 ± 3.8715	0.0394 ± 0.0022	21.813 ± 1.0297
	CD	122.1 ± 3.1691	0.0371 ± 0.0016	23.516 ± 0.8630
P value		0.3124	0.2764	0.1046
Partial length (mm)	ED	37.26 ± 1.0098[a]	0.0415 ± 0.0023	16.465 ± 0.8371
	CD	35.56 ± 0.8304[b]	0.0425 ± 0.0021	15.978 ± 0.7135
P value		0.0199	0.5618	0.4519
Cumulative food intake (g)	ED	15.19 ± 0.6551	0.0482 ± 0.0026	42.563 ± 1.0919
	CD	15.33 ± 0.5732	0.0485 ± 0.0023	42.656 ± 0.9413
P value		0.5828	0.7863	0.5979
Cumulative protein intake (g)	ED	4.56 ± 0.1970[b]	0.0482 ± 0.0026	42.563 ± 1.0919
	CD	5.42 ± 0.5732[a]	0.0485 ± 0.0023	42.655 ± 0.9413
P value		0.0001	0.7863	0.8405
Total body protein (mg)	ED	873.8 ± 0.1837[a]	0.0478 ± 0.0122	43.759 ± 2.3173
	CD	697.0 ± 0.0373[b]	0.0672 ± 0.0062	41.271 ± 1.0896
P value		0.0265	0.0817	0.2525
Total body water (mg)	ED	9.103.8 ± 0.8588[a]	0.0564 ± 0.0088	37.461 ± 2.2084
	CD	8.168.8 ± 0.3603[b]	0.0599 ± 0.0048	36.467 ± 1.0097
P value		0.0028	0.5940	0.1574
Total body fat (mg)	ED	469.4 ± 0.0864	0.0568 ± 0.0154	43.961 ± 3.9850
	CD	421.5 ± 0.0330	0.0592 ± 0.0061	46.103 ± 1.6829
P value		0.6612	0.4787	0.1197
Total body ash (mg)	ED	195.6 ± 0.0444	0.0443 ± 0.0105	48.064 ± 2.932
	CD	169.6 ± 0.0124	0.0528 ± 0.0043	47.024 ± 1.706
P value		0.1044	0.0545	0.7943

Pm = weight or length at maturity; *b* (per day) = maturation rate; t^* (days) = time of maximum growth rate. Means in the same column followed by different superscript letters differ significantly ($P < 0.05$, *F* test).**Source:** Elaboration of the authors. $Wt = sWm \times \exp \times (-\exp \times (-b \times (t - t^*)))$, where *Wt* = nutrient weight (g) of the animal at time *t*, expressed as a function of *Wm*; *Wm* = nutrient weight (g) at maturity of the animal; *b* = maturation rate (per day); t^* = time (days) when the growth rate is maximal. [1]ED = 26.23% digestible protein and 32.68% crude protein; [2]CD = 37.92% crude protein.

Table 1. Parameter estimates obtained with the Gompertz equation for live weight, feed and protein intake, total and partial lengths and nutrient deposition of bullfrog tadpoles fed the experimental (ED)[1] and commercial (CD)[2] diets.

Age (days)	Protein (g/dia)		Lipid		Water	Ash	
	CONS	DEP*	CONS	DEP*	DEP*	CONS	DEP*
1	0.0058	0.00064	0.0008	0.00016	0.0104	0.00109	0.00005
12	0.0155	0.00399	0.0023	0.00128	0.0569	0.00294	0.00053
20	0.0254	0.00796	0.0037	0.00293	0.1069	0.00482	0.00126
29	0.0414	0.01147	0.0061	0.00485	0.1475	0.00787	0.00201
42	0.0537	0.01207	0.0079	0.00600	0.1497	0.01019	0.00226
54	0.0642	0.00947	0.0094	0.00527	0.1155	0.01218	0.00179
63	0.0608	0.00711	0.0089	0.00711	0.0862	0.01155	0.00134

*Deposition values were estimated by the derivative of the Gompertz equation.

Table 2. Consumption (CONS) and deposition (DEP) of nutrients according to age of bullfrog tadpoles.

the diets presented the best performance for the animals. Thus, it was possible to conclude that the Gompertz model provided a good fit of the data to describe the morphometric growth curve and carcass nutrient deposition of bullfrog tadpoles. A higher growth rate and nutrient deposition was observed for tadpoles receiving the experimental diet (26.23% digestible protein).

On the basis of the estimated equation, growth rates (g/day) were calculated as a function of time (t) by the derivative $dWt/dt = bWt\exp(-b(t - t^*))$ of the equation described by Winsor [65].

Still taking as an example the Gompertz equation, the growth, consumption, and nutrient deposition rates (g/day) as a function of time (t) can be calculated by means of the derivative of the equation $dY/dK = bPt\exp(-b(t - t^*))$, Winsor [65]. These parameters are very simple to obtain, an example of which is their estimation by the modified Gauss-Newton method using nonlinear regression using the NLIN procedure of SAS or another statistical program.

It was possible to verify that the values of the consumption, deposition, protein, fat, moisture, and ash content weights showed as the tadpoles gained body protein weight, there was an increase in the deposition of the other nutrients. After deposition of nutrients in the tadpole body, ash, protein, and water deposition occurred in the initial phase (**Table 2**). The authors concluded that nutrient consumption is greater than the nutrient deposition in the carcass of the bullfrog tadpoles (*Lithobates catesbeianus*) and the high protein content of 57.53% of the commercial feed used is not fully utilized by the bullfrog tadpoles.

3. Relative growth and allometric coefficients of body components of fish and amphibian

3.1. Allometric growth

The body composition of the fish changes throughout the life cycle and its utilization is affected by endogenous (species, size) and exogenous factors such as time of year and diet composition [66]. According to Bureau et al. [67], the nutritional factors of the rations such as the balance of available amino acids, essential amino acids, amount of protein and the ratios of protein: energy is important in the deposition of protein and lipid in the tissues. Therefore, during the growth, there are seasonal changes in body composition, associated with the endocrine states and the special physiological stages. At the reproduction stage, there occur the syntheses and reserves of new tissues [26]. In order to analyze this dynamics, the nutrient prediction models can be used. These are mechanistic models which are used to define the destination of dietary nutrients, considering the use of amino acids, fatty acids and their precursors [26]. Thus for example, the amount of protein in the body can be described by means of a growth function. However, the increase in water, ash, and lipid deposition may be linked to protein to determine the whole body growth [39].

3.2. Equations for predicting allometric growth

The isometric and allometric relationships based on regression analysis are still successful to estimate the body composition in fish and other animals in the production sector [26]. The different genotypes may differ in aspects that are estimated from growth curves, such as the maturity, body composition at maturity, fat content, and maturity rates of the body chemical components. The chemical composition varies over time [39]. The energy gain can be predicted using the bioenergetic models, but these do not provide much information on the chemical composition and biomass gain [26].

Allometry refers to changes in the different dimensions of body parts that are correlated with changes in the whole body [68]. According to Thornley and France [34], allometry means growth of a part of the body (W_1) related to a different proportion of the whole body (W). It may be expressed as follows: $y = aX^b$, where a is the normalization constant, b is the dimensions of allometric parameters. This equation can be linearized as follows: $\ln y = \ln a + b \ln X$. When the value of b is equal to 1, the growth is considered isogonic and the rates of development of Y and X are similar in the considered growth interval. In the case of b being greater than 1, the growth is called heterogenic positive and the growth rate of Y is greater than X, characterizing a late development. When the value of b is less than 1, the growth rate of Y is less than X characterizing an early development.

3.3. Allometric evaluation to describe growth variables

As an example, in a study conducted with the freshwater angelfish [69], it was possible to better understand the applicability of allometry. The allometric coefficients for length, weight, protein, fat, ash, and water were determined. The allometric equations and their components in addition to the coefficient of determination (R^2) of the standard length (SL), head length (HL), height (H) and width (W) ratios are shown in **Table 3**. For the height component, the value of b was 1.095 indicating that the fish presented a positive allometric growth or isogonic growth ($b = 1$), that is, from 30 to 233 days of age, the height increased by the same magnitude as the standard length.

Other components, such as the head length and width, showed an early growth ($b < 1$), increased at a lower rate than height, but with more intensity in the final phase of the growth

Components	Coefficients *Ln a*	*B*	R^2
Head length	−0.678	0.907	0.907
Height	−0.728	1.095	0.900
Width	−1.029	0.749	0.886
Weight	−10.25	3.060	0.989

Length (L), width (W), weight and standard length (SL), natural logarithm of the normalization constant (*Ln a*), dimensions of allometric parameters.

Table 3. Allometric coefficients of *P. scalare* juveniles from 30 to 233 days of age in relation to the standard length.

period (**Figure 2**). According to Santos et al. [70], head growth is early to ensure feed consumption during the early stages of fish growth; in late adult years there is a late growth.

In this study, males and females were kept together in the aquariums and it was not possible to estimate the sex ratio due to the difficulty in identifying sexual dimorphism in the early stages of growth. The difference between the growth rates of the different parts of the fish was more noticeable throughout the structuring period and stabilized when they got maturity level.

In production fish, it is useful to know the growth of the fillet in relation to its body weight in order to estimate the possible slaughter weight of the animal. Gomiero et al. [44] evaluated the development of fillet in relation to body weight in Piracanjuba fish, which presented an isogonic growth. According to the results obtained by Almeida et al. [71] about *Oreochromis niloticus* grown in a semi-intensive system, the fillet growth was smaller than the body growth, whereas in an intensive rearing system, the fillet presented an enlargement equal to the body weight with a value of $b = 0.9690$.

Taking as an example, the result of the analysis of the standard length and weight ratio of freshwater angelfish presented in **Figure 3a** and **b**, we can observe that this fish has an isogonic growth indicating a proportional increase in weight and length. In **Figure 3b**, the

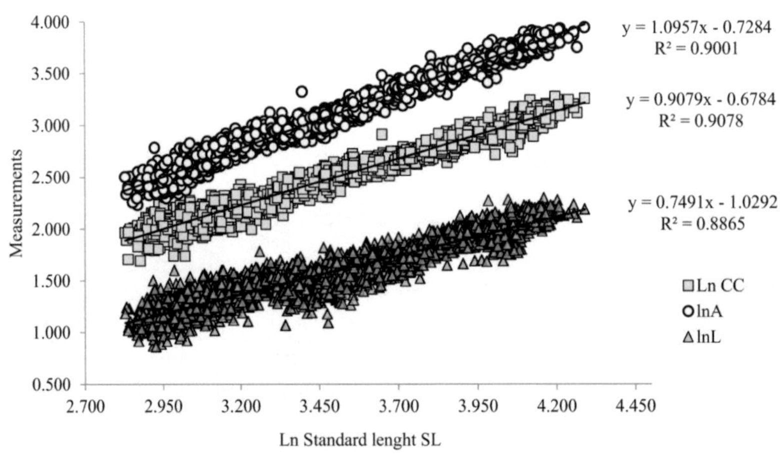

Figure 2. Allometric lengths of head (H), height (H), width (W), and standard length (SL) in *P. scalare* from 30 to 233 days of age.

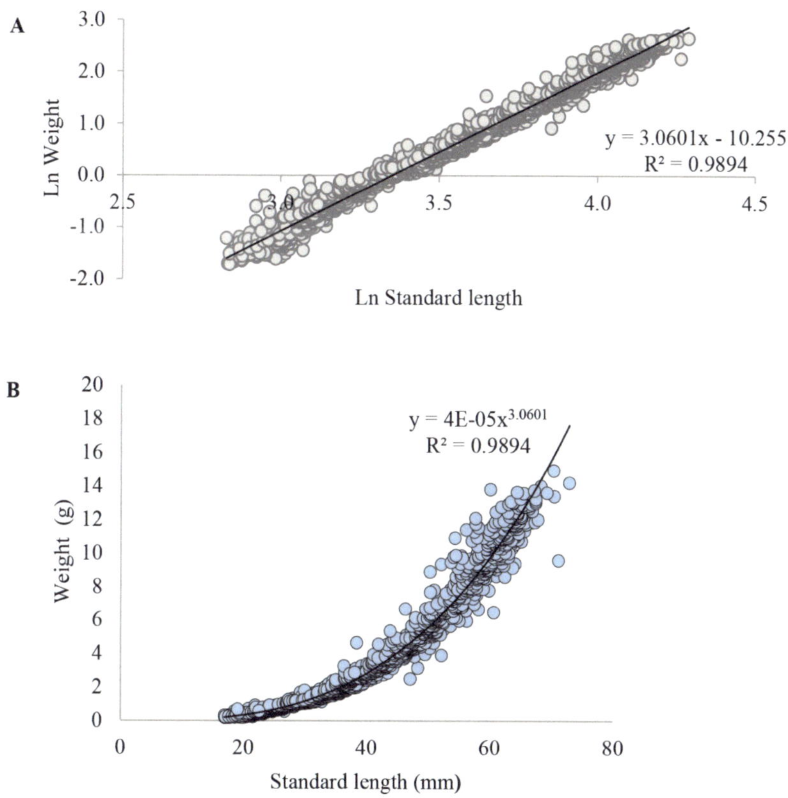

Figure 3. Allometric relationships between weight and standard length in *P. scalare* from 30 to 233 days of age. *A* is the linear regression and *B* is the determination coefficient.

exponential relation between standard length and weight can be observed. The coefficient of determination was 9.089 for the two regressions.

The value of *b* obtained for freshwater angelfish was 3.06 (**Figure 3a**). Results are close to those already found for the same and other fish species. For example, in *Arapaima gigas* grown in a semi-intensive system in the state of Amazonas, Tavares-Dias et al. [72] obtained for the coefficient *b* a value of 3.068. Silva-Júnior et al. [73] obtained a value of *b* between a range of 2.4 and 3.4 for 33 estuarine fish species. Sani et al. [74], studying 14 species of freshwater fish in India, also have found a range of the value of *b* between 2.4 and 3.52. These values are within the ideal for fish, which should be close to 3 [75]. The time of year had influence on the weight-to-length relationship in salmon (*Salmo trutta*); in winter, there was negative allometry and in the other seasons the growth was isometric for females, males, and the mixed group [76]. According to Tavares-Dias et al. [72], knowing the value of body weight can estimate the value of the standard length or vice versa.

Due to the difficulty in identifying the sexual dimorphism in the flag mites at the time of the beginning of the experimental phase, the fishes were not separated by sex, and for the biometric analysis, the fishes were randomly selected. In order to evaluate the length-to-weight relationship in *Buglossidium luteum*, separating the fish in groups by sex and also in mixed groups, İlkyaz et al. [77] concluded that although the length between sexes presented different values, the weight-length curves were very similar thereby growth was isometric for females, males, and for the mixed groups.

Several allometric relationships exist in the literature describing the relationship between surface area (SA) and body mass (BM) for different species of Anurans and these are frequently used in physiological studies. However for species of production such as bullfrog, little studies exist on such allometric relation. In the study conducted by Klein et al. [78], the bibliographic data such as, surface area (SA) (cm^2) and body mas (BM) (g) was collected, and the allometric relationships between SA and BM were evaluated using linear regressions and phylogenetic generalized least squares (PGLS). Data from 453 specimens of 44 species were included. Intraspecific allometric relationships between SA and BM were determined for 18 species, of which 10 presented regressions significantly different from the respective family regression, four species showed a significantly different intercept-*y*, and three species exhibited a significantly different slope. Only the Bufonidae, Ranidae, and Hylidae families were represented by several species (9, 11, and 12, respectively) and with a larger number of specimens (54, 215, and 127, respectively). These three families showed significantly different OLS linear regressions on log-transformed data, with Hylidae being the steepest (0.7735 ± 0.0110), Bufonidae an intermediate (0.6772 ± 0.0220), and Ranidae the lowest slope (0.6091 ± 0.0114). The relationship between SA and BM for Anura could be described by linear regression SA = 9.8537 BM 0.6745 or by the regression of PGLS SA = 8.7498 BM 0.685.

3.4. Allometric evaluation for dynamics of macromolecules

With the allometric equations, it is possible to determine the relationship of body nutrients in relation to protein weight or live weight. Thus, nutrient prediction as a function of protein weight corrects changes in body fat related to diet [79]. The amount of protein may be described as a growth function, and then the growth of water, ashes, and lipids may be linked to protein to determine the growth rate of the whole body [39]. Although the lipid and ash contents separately are not good predictors of body weight [80].

Figure 4 shows the allometric coefficients for the freshwater angelfish body components. The allometric relationship between body protein and live weight showed an isogonic tendency ($b = 1.037$), protein increased in the same proportion as body weight and these observations agree with the study done by Dumas et al. [80] with trout. This can be explained by the fact that the weight of the protein is linked to the live weight [80] mainly by the muscular gain. The fat performance regarding live weight was higher and 1105 units of fat were deposited per unit of live weight. Fat is the most dynamic macromolecule, and its rate of change is easily affected by the temperature of the water in which the fish exist, amount of fat (energy) in the diet, in addition to if the diet that has a protein imbalance. For each component, the coefficient of determination R^2 was above 0.99 presenting a good fit of the model to the data.

In allometric study of Nile tilapia of GIFT strain, due to the body weight, Amancio et al. [81], found that as the fish gained body weight, there was an increase in the proportion of protein ($b = 1.039$), fat ($b = 1.089$) and ash ($b = 1.051$) and a reduction in body water ratio ($b = 0.983$).

http://dx.doi.org/10.5772/intechopen.69061

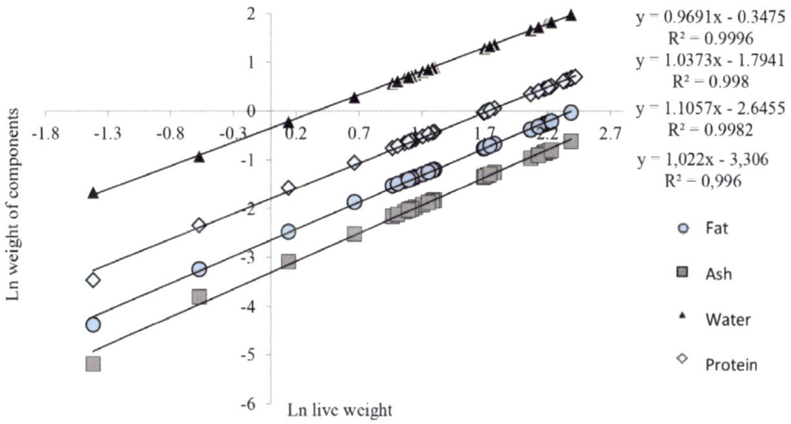

Figure 4. Allometric relationships between live weight and fat, ash, water and protein in *P. scalare* from day 30 to day 233 of age.

This lower water weight ratio may be a result of the increased fat proportions in the carcass, since this was the component that presented the highest allometric coefficient.

In the study conducted by Silva [82], it was verified that in Nile tilapia of supreme strain, the body nutrient that increases largely as body weight increases is fat, mainly to the detriment of moisture content. The inverse relationship between lipid and water contents in the fish muscle was also observed by Guinazi et al. [83], Caula et al. [84], and Neves [85]. Even with the use of allometric equations derivative, it is possible to gain weight of certain nutrients per gram of body weight or protein weight. According to Bureau et al. [67], protein deposition governs the growth of the animal, since for each gram of protein deposit, three and six grams of water deposited, while the deposition of lipids can be done by replacing the water.

Allometric equations are important in determining the relationship of body nutrients, organs, muscle, bone, and skin to protein weight or live weight. The equations can estimate the content of nutrients that the animal deposits are based on the protein weight or live weight [79]. Allometry has been used in mathematical modeling because the body composition of lipid-free dry matter does not change during animal development, but the lipid content of growing animals can be affected by the diet [86].

When allometry has been used in relation to the proportion of protein in the body, the differences between sex and lineages are small. Thus, the use of protein weight in allometric relationships makes the equations more precise. However, the development of allometric equations in relation to fasting live weight would be the most practical method to predict body weight and body nutrient deposition [79].

4. Final considerations

Growth models are useful tools that besides evaluating variables within a population, allow measures to improve the points of the curve, make a selection of the desirable characteristics within a production system and allow improving the feeding strategies for the animals. For future allometric growth research in order to improve the standardization of values, different animal groups can be selected for conducting trials while separating them on the basis of their sex, age, sexual maturity, and the time of year.

Acknowledgements

We thank the state funding agency Fundacão de Amparo à Pesquisa do Estado de São Paulo (FAPESP) for financial support (Grants 2013/25761-4).

Author details

Cleber Fernando M. Mansano[1]*, Beatrice Ingrid Macente[2], Kifayat Ullah Khan[1], Thiago Matias T. do Nascimento[1], Edney P. da Silva[2], Nilva Kazue Sakomura[2] and João Batista K. Fernandes[1]

*Address all correspondence to: clebermansano@yahoo.com.br

1 Aquaculture Center of UNESP, São Paulo State University, Jaboticabal, SP, Brazil

2 Faculty of Agrarian and Technological Sciences, São Paulo State University, Jaboticabal, SP, Brazil

References

[1] Ensminger ME, Oldfield JE, Heinemann WW. Feeds and Nutrition. 2nd ed. The Ensminger Publishing Company. Califórnia; 2001. p. 1544

[2] Gonzales E, Sartori JS. Crescimento e Metabolismo Muscular. Fisiologia Aviária Aplicada a Frangos de Corte. Jaboticabal, SP, Brazil: FUNEP/UNESP; 2002. pp. 279-298

[3] Lawrence TLJ, Fowler VR. Growth of Farm Animals. CAB: New York; 1997. p. 330

[4] Marcato SM, Sakomura NK, Fernandez JBK, Nascimento DCN, Furlan RL, Piva GH. Crescimento e Deposição de Nutrientes nas Penas, Músculo, Ossos e Pele de Frangos de Corte de Duas Linhagens Comerciais. Ciência e Agrotecnologia. 2009;**33**(4):1159-1168

[5] Albinati RCB, Lima SL, Donzele JL. Níveis de Energia Digestível na Ração de Girinos de Rã-touro. Revista Brasileira de Saúde e Produção Animal. 2001;**2**:48-52

[6] Hota AK. Growth in amphibians. Gerontology. 1994;**40**:147-160

[7] Manwell C. Metamorphosis and gene action-i. electrophoresis of dehydrogenases, esterases, phosphatases, hemoglobins and other soluble proteins of tadpole and adult bullfrogs. Comparative Biochemistry and Physiology. 1966;**17**:805-823

[8] Agostinho CA, Silva MA, Torres RA, Lima SL. Parâmetros genéticos de características de produção em rã-pimenta (Leptodactylus labyrinthicus) (Spix, 1824). Revista Sociedade Brasileira de Zootecnia. 1991;**20**:55-60

[9] Mazzini ARA, Muniz JA, Aquino LH, Silva FF. Análise da curva de crescimento de machos hereford. Ciência e Agrotecnologia. 2003;**27**(5):1105-1112

[10] Emmans GC. A model of the growth and feed intake of ad libitum fed animals, particularly poultry. In: Hillyer GM, Whittemore CT, Gunn RG, editors. Computers in Animal Production. 5th ed. British Society of Animal production. Occasional publication; Edinburgh; 1981. pp. 103-110

[11] Leeson S, Summers JD. Commercial Poultry Nutrition. 2nd ed. University Books, Guelph; 1997. p. 355

[12] Gous RM. Modelling energy and amino acid requirements in order to optimize the feeding of commercial broilers. In: 2nd International Symposium on Avian Nutrition; Concordia, Brazil; 2001

[13] Hammond J. Farm Animals. 3rd ed. London: Edward Arnold Publishers Ltd; 1960

[14] Johnston IA. Muscle development and growth: Potential implications for flesh quality in fish. Aquaculture. 1999;**177**:99-115

[15] Gómez DAA, Cerón MFM, Restrepo LFB. Modelación de funciones de crecimiento aplicadas a la producción animal. Revista Colombiana Ciencias Pecuarias. 2008;**21**:39-58

[16] Altig R, McDiarmid RW. Body plan: developmental and morphology. In: McDiarmid RW, Altig R, editors. Tadpoles: The Biology of Anuran Larvae. Chicago: Chicago Press; 1999. pp. 24-51

[17] Gosner KL. A simplified table for staging anuran embryos and larvae with notes on identification. Herpetologica. 1960;**16**:183-190

[18] Reeder WG. The digestive system. In: Moore JA, editor. Physiology of the Amphibia. Vol. 1. New York: Academic Press; 1964;**1**:654

[19] Werner EE, Willborn GA, McPeek MA. Diet composition in post metamorphic bullfrog and green frogs: Implications for interspecific predation and composition. Journal of Herpetology. 1995;**29**(4):600-607

[20] Hirai T. Diet composition of introduced bullfrog (*Rana catesbeiana*), in the Mizorogaike Pond of Kyoto, Japan. Ecological Research. 2004;**19**:375-380

[21] Silva ET, Reis EP, Feio RN, Ribeiro OPF. Diet of the invasive frog (*Lithobates catesbeianus*) (Shaw, 1802) (Anura: Ranidae) in Viçosa, Minas Gerias State, Brazil. South American Journal of Herpetology. 2009;**4**(3):286-294

[22] Hepher B. Growth. In: Hepher B, editor. Nutrition of Pond Fishes. Cambridge: Cambridge University; 1993. pp. 163-191

[23] Barbato GF, Vasilatos-Younken R. Sex-linked and maternal effects on growth in chickens. Poultry Science. 1991;**70**:709-718

[24] Dutta H. Growth in fishes. Gerontology (India). 1994; **40**:97-112

[25] Ramos S. Ajustes de curvas de crescimento e estimativas da variabilidade genética de peso corporal de avestruzes (*Struthio camelus*). 2010. 48 f. Dissertação (Mestrado em Genética e Melhoramento Animal). Jaboticabal, SP, Brazil: Faculdade de Ciências Agrárias e Veterinárias, Universidade Estadual Paulista "Julio De Mesquita Filho"; 2010

[26] Dumas A, France J, Bureau D. Modelling growth and body composition in fish nutrition: where have we been and where are we going? Aquaculture Research. 2010;**41**: 161-181

[27] Borges FF, Amaral LA, Stefani MV. Characterization of effluents from bullfrog (*Lithobates catesbeianus*, Shaw, 1802) grow-out ponds. Acta Limnologica Brasiliensia. 2012;**24**(2):160-166

[28] Teodoro SM, Chaves MA, Escobedo JF, Agostinho CA. Relação de variáveis ambientais em baias cobertas com polietileno e desempenho da rã-touro (*Rana catesbeiana*). Engenharia Agrícola. 2005;**25**(1):46-56

[29] Petersen AM, Gleeson TT. Acclimation temperature affects the metabolic response of amphibians skeletal muscle to insulin. Comparative Biochemistry and Physiology Part A. 2011;**160**:72-80

[30] Braga SL, Lima SL. Influência da temperatura ambiente no desempenho da rã-touro, *Rana catesbeiana* (Shaw, 1802) na fase de recria. Revista Brasileira de Zootecnia. 2001;**30**(6):659-1663

[31] Figueiredo MRC, Agostinho CA, Baêta FC, Lima CA. Efeito da temperatura sobre o desempenho da rã-touro (Rana catesbeiana, Shaw 1802). Revista Brasileira de Zootecnia. 1999;**28**(4):661-667

[32] Figueiredo MRC, Lima SL, Agostinho CA, Baêta FC. Efeito da temperatura e do fotoperíodo sobre o desenvolvimento do aparelho reprodutor de rã-touro (*Rana catesbeiana* Shaw, 1802). Revista Brasileira de Zootecnia. 2001;**30**:916-923

[33] Tedeschi LO. Review assessment of the adequacy of mathematical models. Agricultural Systems. 2006;**89**:225-247

[34] Thornley JHM, France J. Mathematical Models in Agriculture: Quantitative Methods for the Plant, Animal and Ecological Sciences. 2nd ed. Wallingford: CABI; 2007. p. 906

[35] Martínez R, Martínez N. Diseño de experimentos: Análisis de datos estándar y no estándar. Bogotá: Editorial Guadalupe; 1997. p. 479

[36] Rondón-oviedo O, Waldroup PW. Models to estimate amino acid requirements for broiler chickens: A review. International Journal of Poultry Science. 2002;**1**(5):106-113

[37] Parks JA. Theory of Feeding and Growth of Animals. Berlin: Springer-Verlag; 1982. p. 451

[38] Strathe AB, Danfaer A, Sørensen H, Kebreab EA. A multilevel nonlinear mixed effects approach to model growth in pigs. Journal of Animal Science. 2010;**88**:638-649

[39] Gous RM, Moran Jr, ET, Stilborn HR, Bradford GD, Emmans GC. Evaluation of the parameters needed to describe the overall growth, the chemical growth, and the growth of feathers and breast muscles of broilers. Poultry Science. 1999;**78**:812-882

[40] Brown D, Rothery P. Models in Biology: Mathematics, Statistics and Computing. England, Chichester, New York, Brisbane, Toronto, Singapore: John Wiley and Sons; 1993. p. 688

[41] Brown JE, Fitzhugh Jr. HA, Cartwright TCA. Comparison of nonlinear models for describing weight-age relationships in Cattle. Journal of Animal Science. 1976;**42**:810-818

[42] Barlow J. Nonlinear and logistic growth in experimental populations of Guppies. Ecology. 1992;**73**(3):941-950

[43] Keele JW, Williams CB, Bennett GL. A computer model to predict the effects of level of nutrition on composition of empty body gain in beef cattle. I. Theory and development. Journal of Animal Science. 1992;**70**(3):841-857

[44] Gomiero JSG, Freitas RTF, Santos VB, Silva FF, Rodrigues PB, Logato, PVR. Curvas de crescimento morfométrico de Piracanjuba (*Brycon orbignyanus*). Ciência e Agrotecnologia. 2009;**33**:882-889

[45] Santos VB, de Freitas RTF, Silva FF, Freato TA. Avaliação de curvas de crescimento morfométrico de linhagens de tilápia do nilo (*Oreochromis niloticus*) Ciências. Agrotecnología. 2007;**31**(5):1486-1492

[46] Oliveira HN, Lôbo RB, Pereira CS. Comparação de modelos nãolineares para descrever o crescimento de fêmeas da raça guzerá. Pesquisa Agropecuária Brasileira. 2000;**35**(9): 1843-1851

[47] Silva NAM, Lana AMQ, Silva FF, Silveira FG, Bergmann JAG, Silva MA, Toral FLB. Seleção e classificação multivariada de modelos de crescimento não lineares para bovinos nelore. Arquivos Brasileiro de Medicina Veterinária e Zootecnia. 2011;**63**(2):364-371

[48] Silveira FG, Silva FF, Carneiro PLS, Malhado CHM. Classificação multivariada de modelos de crescimento para grupos genéticos de ovinos de corte. Revista Brasileira de Saúde e Produção Animal. 2011;**13**(1):62-73

[49] Sarmento JLR, Rezazzi AJ, Souza WH, Torres RA, Breda FC, Menezes GRO. Analysis of the growth curve of Santa Ines sheep. Revista Brasileira de Zootecnia. 2006;**35**:435-442

[50] Mansano CFM, Stéfani MV, Pereira MM, Macente BI. Non-linear growth models for bullfrog tadpoles. Ciência e Agrotecnologia. 2012;**36**(4):454-462

[51] Rodrigues ML, Lima SL, Moura OM, Agostinho CA, Silva JHV, Cruz GRB, Campos VM, Casali AP, Mendes RRB, Albuquerque AG. Curva de crescimento em rã-touro na fase de recria. Archivos de Zootecnia. 2007;**56**(214):125-136

[52] Silva FF, Aquino LH, Oliveira AIG. Estimativas de parâmetros genéticos de curva de crescimetno de gado nelore (*Bos indicus*). Ciência e Agrotecnologia. Edição Especial: 2002;1562-1567

[53] Mendes PN, Muniz JA, Silva FF, Mazzini ARA, Silva NAM. Análise da curva de crescimento difásica de fêmeas hereford por meio da função não linear de Gompertz. Ciência Animal Brasileira. 2009;**10**(2):454-461

[54] Amancio ALL, Silva JHV, Fernandes JBK, Sakomura NK, Cruz GRB. Use of mathematical models in the study of bodily growth in GIFT strain Nile tilapia. Revista Ciência Agronômica. 2014;**45**:257-266

[55] Carvalho JC. Desempenho zootécnico e curvas de crescimento de tilápia do nilo (*Oreochromis niloticus*) melhoradas geneticamente para ganho em peso. 2016. 49 f. Dissertação (Mestrado em Ciência Animal), Campo Grande, Brazil: Universidade Federal de Mato Grosso do Sul; 2016

[56] Hernandez-Llamas F, Ratkowsky DA. Growth of fishes, crustaceans and molluscs: Estimation of the von Bertalanffy, Logistic, Gompertz and Richards curves and a new growth model. Marine Ecology Progress Series. 2004;**282**:237-244

[57] Katsanevakis S, Maravelias CD. Modelling fish growth: Multi-model inference as a better alternative to a priori using von Bertalanffy equation. Fish and Fisheries. 2008;**9**(2):178-187

[58] Aguilar FA. Modelos matemáticos no lineales como herramienta para evaluar el crecimiento de tilapia roja (*Oreochromis* spp.) tilapia nilótica (*Oreochromis niloticus* Var. Chitralada)" alimentadas con dietas peletizadas o extruidas. 135 f. Dissertação (Mestrado em Produção Animal). Bogotá: Faculdade de Medicina Veterinária e de Zootecnia, Universidade Nacional de Colômbia; 2010

[59] Costa AC, Neto Reis RV, Freitas RTF, Freato TA, Lago AA, Santos VB. Avaliação do crescimento de tilápias de diferentes linhagens através dos modelos não lineares. Archivos de Zootecnia, Córdoba. 2009;**58**:561-564

[60] Silva FL, et al. Growth curves in beef cows of diferente biological types. Pesquisa Agropecuária Brasileira, Brasília. 2011;**46**(3):262-271

[61] Freitas AR. Curvas de crescimento na produção animal. Revista Brasileira de Zootecnia, Viçosa. 2005;**34**(3):786-795

[62] Pereira MM, Mansano CFM, Silva EP, de Stéfani MV. Growth in weight and of some tissues in the bullfrog: Fitting nonlinear models during the fattening phase. Ciência Agrotecnologia, Lavras. 2014;**38**(6):598-606

[63] Gamito S. Growth models and their use in ecological modelling: An application to a fish population. Ecological Modelling. 1998;**113**:83-94

[64] Mansano CFM, de Stéfani MV, Pereira MM, Nascimento TSR, Macente BI. Morphometric growth characteristics and body composition of bullfrog tadpoles in captivity. Semina. 2014;**35**(5):2817-2830

[65] Winsor CP. The Gompertz curve as a growth curve. Proceedings of the National Academy of Sciences of the United States of America. 1932;**18**(1):1-8

[66] Bureau DP, Kaunshik SJ, Cho CY. Bioenergetics. In: Halver JE, Hardy RW, editors. Fish Nutrition. 3rd ed. San Diego, CA, USA: Academic Press; 2002. pp. 1-59

[67] Bureau DP, Azevedo PA, Tapia-salazar M, Cuzon G. Pattern and Cost of growth and nutrient deposition in fish and shrimp: Potencial implications and applications. In: Cruz-suárez LE, Ricque-marie D, Tapia-salazar M, Olvera-novoa MA, Civera-cerecedo R, editors. Avances en Nutrición Acuícola: Simposium Internacional de Nutrición Aacuícola. Mérida. Anais. Mérida, Yucatán, México; 2000. pp. 119-122

[68] Gayon J. History of the concept of allometry. American Zoologist. 2000;**40**:748-758

[69] Manrinque CHE, Fernandes JBK, Sakomura NK, Vigoya AAA, Nascimento TMT, Silva EP, Mansano CFM. Description of growth and body composition of freshwater Angelfish (*Pterophyllum scalare*) by Gompertz model. Revista Brasileira de Zootecnia. 2017x. In Press

[70] Santos VB, Freato TA, Freitas RTF, Logato PVR. Crescimento relativo e coeficientes alométricos de componentes do corpo de linhagens de Tilápias-do-Nilo (*Oreochromis niloticus*). Ciência Animal Brasileira. 2006;**7**(4):357-364

[71] Almeida AK, Ferreira TC, Silva SLH, Kuster LDS, Pires AV, Pereira IG, Júnior FFS. Alometria do crescimento do filé de tilápia (*Oreochromis niloticus*) em dois sistemas de produção. ZOOTEC: 22-26 May, 2006; Centro de Convenções de Pernambuco, Brazil

[72] Tavares-dias M, Barcellos JFM, Marcon JL, Menezes GC, Ono EA, Affonso EG. Hematological and biochemical parameters for the pirarucu (*Arapaima gigas* Schinz, 1822) (*Osteoglossiformes, Arapaimatidae*) in net Cage culture. Journal of Applied Ichthyology. (Berlin). 2007;**2**:61-68

[73] Silva-júnior MG, Castro ACL, Soares LS, França VL. Relação peso-comprimento de espécies de peixes do estuário do rio Paciência da Ilha do Maranhão, Brasil. Boletin do Laborátorio de Hidrobiologia. 2007;**20**:31-38

[74] Sani BK, Gupta UK, Sarkar A, Pandey V, Dubey VK, Singh WL. Length–weight relationships of 14 Indian freshwater fish species from the Betwa (Yamuna River tributary) and Gomti (Ganga River tributary) rivers. Technical note. Journal of Applied Ichthyology. 2010;**26**:456-459

[75] Hile R. Summary of investigations on the morphometry of the cisco (*Leucichthys artedi*) (Le Sueur), in the Lakes of the Northern Highland, Wisconsin. Papers of the Michigan Academy of Science, Arts, and Letters. 1936;**21**:619-634

[76] Arslan M, Yildirim A, Bekta S. Length-weight relationship of brown trout (*Salmo trutta* L.), inhabiting Kan Stream, çoruh Basin, North-Eastern Turkey. Turkish Journal of Fisheries and Aquatic Sciences. 2004;**4**:45-48

[77] İlkyaz TA, Metin G, Soykan O, Kinacigil T. Age, growth and sexual development of solenette (*Buglossidium luteum*) (Risso, 1810), in the central Aegean Sea. Journal of Applied Ichthyology. 2010;**26**:436-440

[78] Klein W, Dabésc, L, Bonfim VMG, Magrini L, Napoli MF. Allometric relationships between cutaneous surface area and body mass in anuran amphibians. Zoologischer Anzeiger. 2016;**263**:45-54

[79] Marcato SM, Sakomura NK, Munari DP, Fernandes JBK, Kawauchi IM, Bonato MA. Growth and body nutrient deposition of two broiler commercial genetic lines. Brazilian Journal of Poultry Science. 2008;**10**:117-123

[80] Dumas A, Lange CFM, France J, Bureau D. Quantitative description of body composition and rates of nutrient deposition in rainbow trout (*Oncorhynchus mykiss*). Aquaculture. 2007;**273**:165-181

[81] Amancio ALL. Características de crescimento e composição corporal da tilápia nilótica (*Oreochromis niloticus*) linhagem gift [Tese (Doutorado em Zootecnia)]. Areia: Universidade Federal da Paraíba; 2011. 91p

[82] Silva TSC. Modelos de crescimento e desempenho de tilápias-do-Nilo (*Oreochromis niloticus*; linhagem Supreme) alimentadas com dietas sem ou com suplementação de lisina e treonina, em gaiolas [Dissertação (Mestrado em Zootecnia)]. Maringá: Universidade Estadual de Maringá; 2008. 50p

[83] Guinazi M, Moreira APB, Salaro AS, Castro FAF, Dadalto M, Pinheiro-sant'ana HM. Composição química de peixes de água doce frescos e estocados sob congelamento. Acta Scientiarum Technology, Maringá. 2006;**28**(2):119-124

[84] Caula FCB, Oliveira MP, Maia EL. Teor de colesterol e composição centesimal de algumas espécies de peixes do estado do Ceará. Ciência e Tecnologia de Alimentos, Campinas. 2008;**28**(4):959-963

[85] Neves HCN. Avaliação do cultivo e das caracteristicas fisico-química, microbiológica e sensorial de tilápia nilótica (*Oreochromis niloticus*) criada no açude Jandaia-Bananeiras/PB. Monografia (Graduação em Agroindústria). Bananeiras: Universidade Federal da Paraíba; 2009. 47p

[86] Vargas GD'A. Modelo de simulação do crescimento e desenvolvimento de frangos de corte [Tese (Doutorado em Produção Animal)]. Pelotas: Faculdade de Agronomia Eliseu Maciel, Universidade Federal de Pelotas; 2004. 116 f

Section 2

Health Sciences

Chapter 3

Morphometrics in Developmental Neurobiology: Quantitative Analysis of Growth Cone Motility *in Vivo*

Anokh Sohal, James Ha, Manuel Zhu,
Fayha Lakhani, Kavitha Thiagaragan,
Lauren Olzewski, Raagav Monakrishnan and
Tamira Elul

Additional information is available at the end of the chapter

http://dx.doi.org/10.5772/intechopen.69060

Abstract

In order for the nervous system to function properly, neurons in the brain must establish specific connections during embryonic development. Formation of neuronal circuits involves axons extending from cell bodies and navigating through diverse tissues to reach their targets in the brain. Once axons reach their target tissues, they arborize and make synaptic connections. Axon pathfinding is driven by dynamic motility behaviors expressed by terminal growth cones at the tips of the axons. Here, we applied morphometrics to determine quantitative values for six morphological and motility parameters for growth cones of optic axons navigating through the optic tract of a living brain preparation from a *Xenopus laevis* tadpole. Our results demonstrate an increase in length, decrease in width, increase in perimeter, decrease in area, increase in number of filopodia, and a decrease in number of lamellipodia, of the growth cones in the optic tract. Therefore, optic axonal growth cones become less circular and more elongated and protrusive during their navigation through the optic tract. Quantitatively deconstructing parameters of growth cone motility is necessary to determine molecular, cellular, and biophysical mechanisms of axon pathfinding, and to formulate computational analyses of developing neuronal connectivity in the brain.

Keywords: morphometrics, axon, growth cone, optic axons, filopodia, lamellipodia, optic tract, *Xenopus laevis*

1. Introduction

In order for the nervous system to function properly, accurate wiring of the brain must be established during embryonic development. Wiring of the brain depends in large part on axons extending from neuronal cell bodies and subsequently navigating through tissues to reach appropriate targets in the brain. Axon pathfinding is driven by specific and patterned motility behaviors expressed by growth cones at the terminal ends of the axons. Quantitatively deconstructing distinct motility parameters of growth cones will aid studies exploring the molecular and mechanical control of axonal pathfinding, as well as facilitate the development of computational analyses of growth cone motility. Here, we have applied morphometric analyses to determine values, and spatiotemporal patterns in those values, for six parameters of motility from a time-lapse video of two growth cones of optic axons navigating through the optic tract of a living brain preparation from a young *Xenopus laevis* tadpole.

During embryonic development of the visual system, optic neurons extend axons from the eye to the tectal midbrain, where they make synaptic connections essential for visual function. The ability of these optic axons to navigate and propel through the optic tract, and to eventually reach the tectum, is due to the growth cone of the axon. The growth cone is a highly motile structure located at the distal end of the axon that mediates its directional growth and extension by interacting with molecular and mechanical cues in the environment. The growth cone can be divided into three sub-compartments, the peripheral (P), transitional (T), and central (C) domains (**Figure 1**). The P-region of the growth cone contains a meshwork of

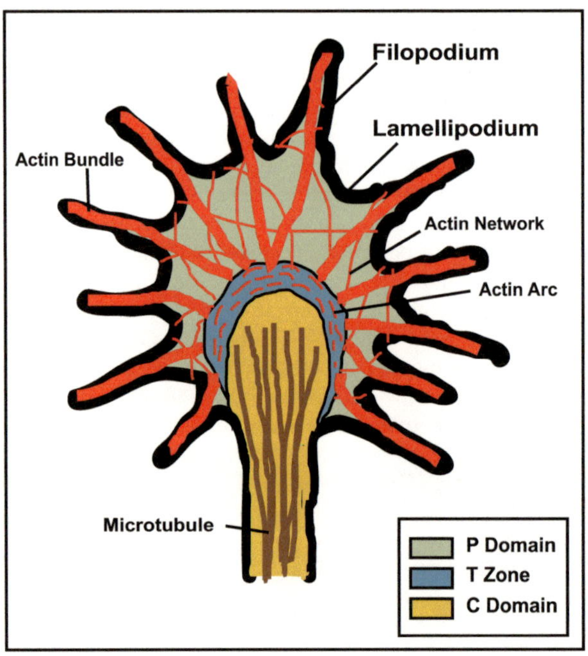

Figure 1. Schematic of growth cone with intracellular domains and protrusions demarcated.

actin filaments and long parallel bundles of actin filaments that underlie two types of protrusions (**Figure 1**). Lamellipodia are short and broad protrusions that are thought to function to generate force for the growth cone advance (**Figure 1**). Filopodia are long and finger-like, and primarily sense the environment and guide the axon (**Figure 1**; [1]). A combination of actin treadmilling and retrograde actin flow allows for continual remodeling of the P-region (and of the lamellipodia and filopodia within this region), required to generate growth cone motility. ATP-actin is assembled into filaments in the distal P-region and then transported rearward to the T-region as polymeric F-actin. In the T-region, F-actin is polymerized and recycled back to ATP-actin and the cycle restarts. Actin is transported retrograde from the P region to the T-region via a myosin motor driven process [2]. The C-region, which is proximal to the P-region, is filled with a dense microtubule array and cellular organelles like mitochondria that support growth cone movement (**Figure 1**; [1]). The microtubule system within the C-region affects cell motility by steering growth cone advance in response to guidance cues from the P-region [3]. The plus end of microtubules exhibits dynamic instability, cycling through periods of growth and shrinkage, allowing them to probe the intracellular space [2]. Similar to actin, microtubules are involved in a transport mechanism involved in maintaining the axon and the growth cone. The majority of microtubules are found in the axon shaft and are stationary. However, in active regions like the growth cone, stable microtubules are tyrosinated to become dynamic [2].

As the growth cone progresses, the P-region senses changes in the extracellular environment, and relays those cues to the C-region. These cues can be either attractive or repulsive [4]. Although the majority of microtubules end in the C-region, single microtubules protrude into the filopodia of the P-region, mediating interaction between the actin and microtubule cytoskeleton (**Figure 1**; [1]). Interactions between microtubules and actin in filopodia are necessary for growth cones to turn. Microtubule and actin interactions also occur in the T zone and C domain of the growth cone, where actin arcs in the T zone exert compressive forces on microtubules in the C domain, facilitating microtubule bundling and aiding in axon navigation (**Figure 1**; [5]). Previous and current studies from our laboratory show that molecules downstream of Cadherin and Wnt signaling ligands such as β-catenin and APC, that regulate the actin and microtubule cytoskeleton, modulate optic axon growth cone morphology and motility in the optic tract as well as targeting and branching in the optic tectum [6, 7]. More generally, it is now clear that the actin and microtubule cytoskeleton of the two growth cone regions are dynamically related, and may influence each other via signaling molecules such as APC [8]. The functional studies of APC *in vitro* from other laboratories also suggest that APC is crucial for growth cone advance and turning [5].

In order to gain a better understanding of the motility, and morphological dynamics of the growth cone, we quantitatively analyzed an *in vivo* time-lapse image sequence of retinotectal axon pathfinding in living brains from *X. laevis* tadpoles. Initial observation of the time-lapse sequence showed that optic axonal growth cones change their morphology and motility depending on where they are in the retinotectal circuit (**Figure 2**). Conceptually, progressing through the optic tract, the growth cone requires greater propulsive properties to get to the tectum but as it reaches the tectum, these propulsive properties decrease as the goal becomes to start making branches and individual synapses with target neurons. We can better understand

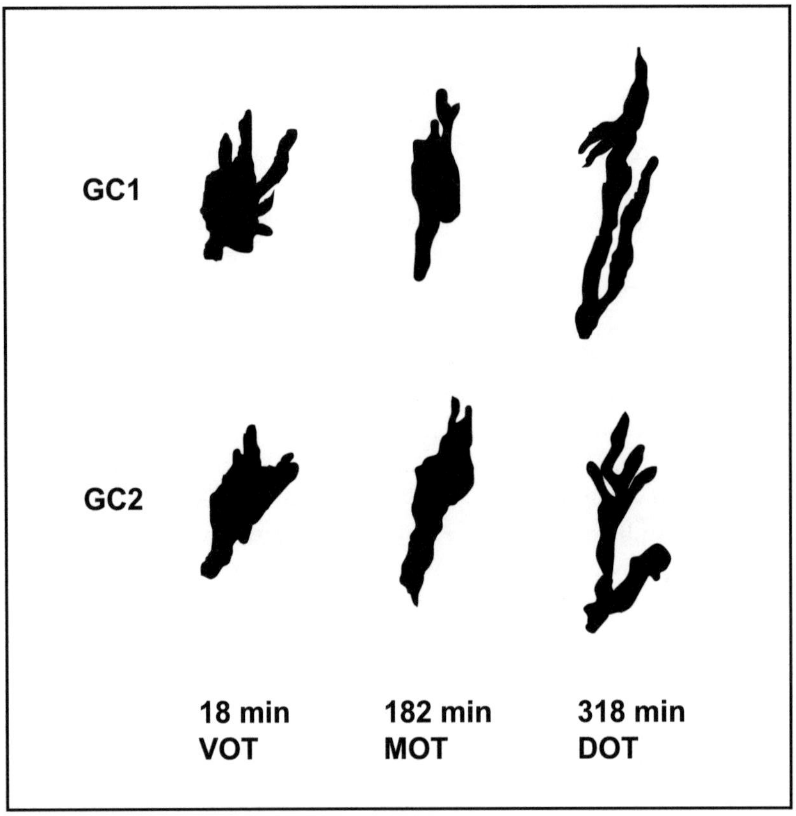

Figure 2. Tracings of still images of two optic axonal growth cones in the optic tract taken from a time-lapse video illustrates how their morphology changes as they navigate through the optic tract. GC1: growth cone 1; GC2: growth cone 2; VOT: ventral optic tract; MOT: mid optic tract; DOT: dorsal optic tract.

how the growth cone behaves and its morphology changes as it progresses through the retinotectal circuit by measuring parameters such as length and width, perimeter and area, and number of filopodial and lamellipodial projections at each time interval [9]. Measurements of length and width, and of perimeter and area, of the growth cone could reflect the details of how the dynamics of actin and microtubule cytoskeleton interactions change as the axon progresses from navigation and propulsion to synapsing. These morphological observations could also help better explain functions of molecular mechanisms such as microtubule tyrosination on the shape of the growth cone. In addition, quantifying the number of filopodial and lamellipodial projections on the growth cones could suggest information on the types of extracellular cues the growth cone is responding to, and their effects on branching of filopodia as the growth cone nears the tectum [9]. Also, observing and quantifying filopodial and lamellipodial morphology could potentially aid researchers to better understand the retrograde actin flow and how it links to the microtubule cytoskeleton through intracellular signaling. Observing and analyzing these different morphological parameters can lead to a better understanding of how intracellular signaling molecules such as β-catenin and APC, or mechanical cues such as tension, affect growth cone motility as well.

http://dx.doi.org/10.5772/intechopen.69060

2. Methods

An *in vivo* time-lapse video of *X. laevis* optic axons navigating in the optic tract and dorsal tectum was made by Sonia Witte in Christine Holt's laboratory at Cambridge University (http://www.pdn.cam.ac.uk/directory/christine-holt; also see Ref. [10]). The video was a collection of 119 individual frames captured at 3-min time intervals of two GFP-expressing growth cones of optic axons navigating through the optic tract of a living brain preparation taken from a young tadpole at approximately developmental stage 33/34 (**Figure 3**; [11]). The time-lapse video sequence encompassed an approximately 6-h (357 min) time period during which two growth cones (labeled with membrane bound GFP) navigate through the optic tract. Analysis of the two growth cones in each frame was performed using the image analysis software ImageJ (NIH, Bethesda, Maryland, USA). Before beginning the analysis, criteria were

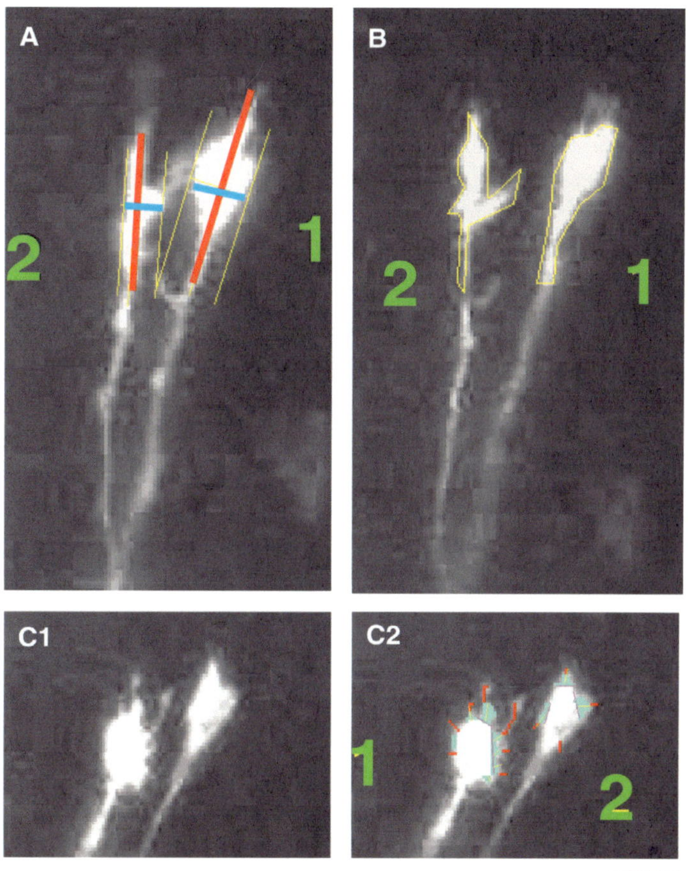

Figure 3. Still images from the time-lapse video sequence of two GFP-labeled optic axons in a living brain preparation from a young *Xenopus laevis* tadpole. (A) Marking for length (thick vertical line) and width (thick horizontal line) measurements of growth cones. (B) Outlines of growth cones for perimeter and area measurements. (C1) Growth cones with protrusions evident. (C2) Filopodia (thin lines) and lamellipodia (thick lines) are marked in the growth cones. Numbers indicate growth cones 1 and 2. Scale Bar—10 μm.

established to standardize the measurements of growth cone parameters (**Figure 3**). A scale bar was set based on a previous publication showing growth cones of optic axons in living brain preparations from *X. laevis* embryos [10].

The length of each growth cone was determined by drawing a line extending along the axonal axis, from the base to the leading edge of the growth cone (thick vertical lines, **Figure 3A**). The base of the growth cone was defined as the first protrusion of the growth cone near the axon shaft (thick lines, **Figure 3A**; [9]). The leading edge of the growth cone was established as the tip of the growth cone, including all protrusions. Width of the growth cone was measured by creating two parallel lines to the length line, at the tips of the distal edges of the growth cone (including its protrusions) (thin vertical lines, **Figure 3A**). The perpendicular distance between these two parallel lines was measured as the maximal width of the growth cone (thick horizontal lines, **Figure 3A**; [9]). This was done for both growth cones for each frame of the time-lapse sequence.

To measure the perimeter of the growth cones, the growth cone boundaries were traced using the polygon drawing tool in ImageJ (yellow outlines, **Figure 3B**). The base of the growth cone was used as the starting point, and the distal edges of the growth cone were traced until reaching the starting point. Similar to the measurement of growth cone length, the base of the growth cone was determined to be where the first protrusion of the growth cone near the axon shaft was located and the growth cone boundary contained all protrusions of the growth cone (**Figure 3B**). Filled area was calculated using the measurement tool in ImageJ.

The number of filopodia and lamellipodia in each image was measured using ImageJ as well (thin lines- filopodia, thick lines - lamellipodia, in **Figure 3C**). Criteria for identifying filopodia and lamellipodia were based on a previous review studying cellular protrusions *in vitro* [12]. This report stated that lamellipodia can vary from 1 to 5 μm in width. Therefore, any protrusions extending from the growth cone body between 1 and 5 μm in width were considered lamellipodium, and any protrusions less than 1 μm in width were considered filopodia. The number of filopodial and lamellipodial protrusions was recorded for both growth cones in each frame.

To avoid researcher-dependent bias in morphometric measurements, five different researchers performed the measurements for length, width, perimeter, and area on the two growth cones at each time point, using the protocols described above. Their values were averaged to obtain final measurements for these size parameters for the growth cones at each time point. All measurements (length, width, area, perimeter, and number of filopodia and lamellipodia) were initially plotted against time in a scatter plot. However, to better display the changes in the data, and to depict the changes in the growth cones as they progressed from ventral, to mid, to finally, the dorsal optic tract, we subdivided the time-lapse video into three time bins. The time-lapse video was composed of 119 frames at 3-min intervals, spanning a total time of 357 min. Therefore, first time bin encompassed 3–117 min (39 images), the second 120–237 min (39 images), and the third 240–357 min (39 images). Averages were calculated for the morphometric measurements for each of the three time groups for both growth cones, and then averages of those averages were determined over the two growth cones. These average growth cone parameters were plotted on bar graphs to determine if there were any differences between their values during the three time bins (**Figures 4–6**).

http://dx.doi.org/10.5772/intechopen.69060

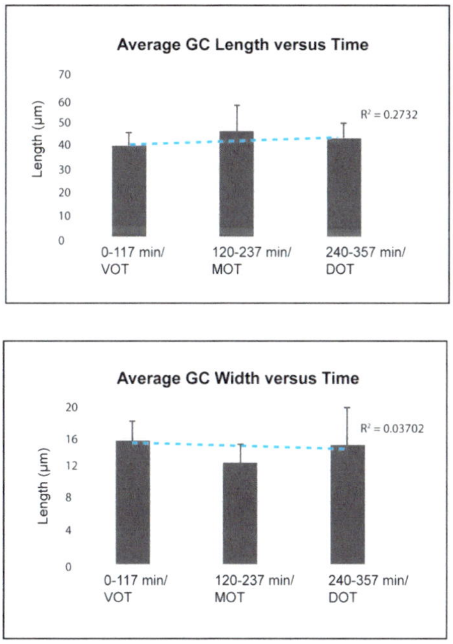

Figure 4. Plots of average growth cone length and width versus time. VOT: ventral optic tract; MOT: mid optic tract; DOT: dorsal optic tract.

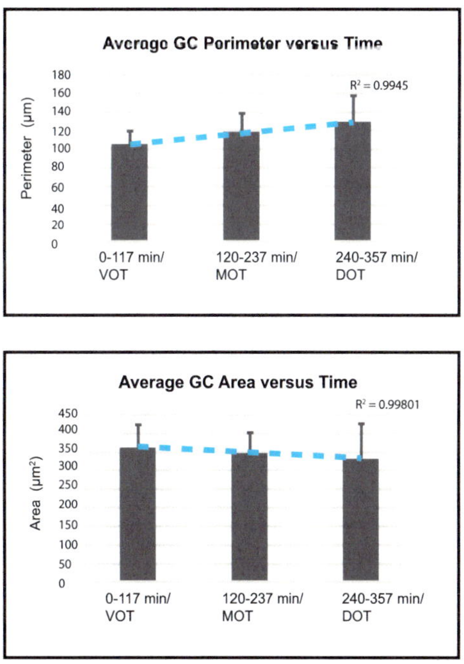

Figure 5. Plots of average growth cone perimeter and area versus time. VOT: ventral optic tract; MOT: mid optic tract; DOT: dorsal optic tract.

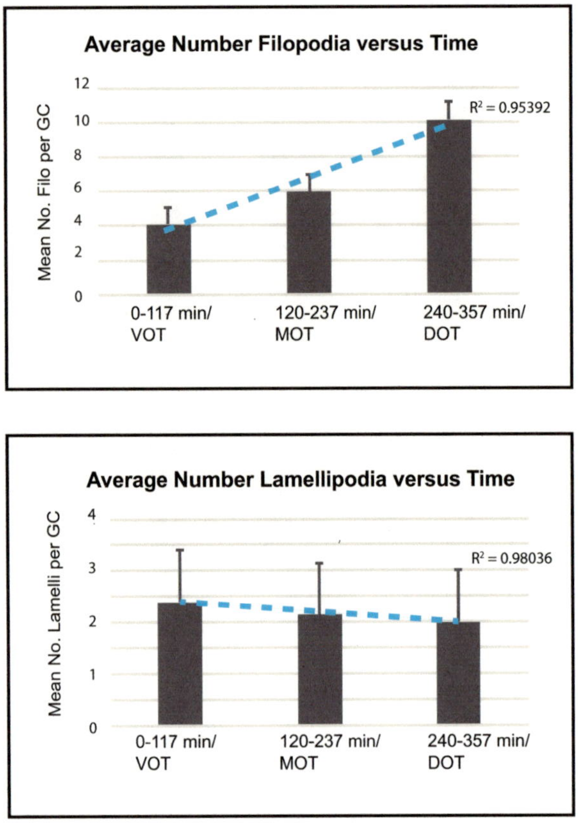

Figure 6. Plots of mean number of filopodia and lamellipodia per growth cone versus time. VOT: ventral optic tract; MOT: mid optic tract; DOT: dorsal optic tract.

3. Results

3.1. Quantitative analysis shows morphological changes in growth cones *in vivo*

3.1.1. Growth cone length increases, while width decreases over time

The lengths and widths of the growth cones were measured using specific criteria described in Section 2. Measurements were taken for both growth cones in each of the 119 frames of the time-lapse video. The time-lapse video was broken up into three equal time bins, and the average length and width for the growth cones were calculated for each of the time bins and plotted on bar graphs (**Figure 4**). Trend lines were added to the graphs. The results revealed that as the growth cones progressed through the optic tract, on average, their length increased, and their width decreased (**Figure 4**). However, the length of the growth cones increased a smaller amount than their width decreased during their navigation through the optic tract (**Figure 4**).

The mean length for growth cone one was initially 45.9 μm (SD = 2.52 μm, n = 39 images), in the ventral optic tract/time bin one. The mean length for growth cone one increased to

47.9 μm (*SD* = 0.78 μm, *n* = 39 images), in the third time bin, corresponding to the dorsal optic tract. The mean length for growth cone two was initially 35.7 μm (*SD* = 8.8 μm, *n* = 39 images), in the ventral optic tract/time bin one. The mean length for growth cone two increased to 40 μm (*SD* = 12.04 μm, *n* = 39 images), in the third time bin, or the dorsal optic tract. On average, the two growth cones increased their lengths by 8% (*SD* = 5.5%, *n* = two growth cones) as they progressed from the ventral to the dorsal optic tract. This corresponds to an average rate of increase of length of 1.4%/h (*SD* = 0.92%/h, *n* = two growth cones) for growth cones of optic axons navigating in the optic tract.

The mean width for growth cone one was initially 16.1 μm (*SD* = 1.81 μm, *n* = 39 images), in time bin one, or the ventral optic tract. The mean width for growth cone one changed to 13.9 μm (*SD* = 3.61 μm, *n* = 39 images), in the third time bin, corresponding to the dorsal optic tract. The mean width for growth cone two was initially 15.3 μm (*SD* = 3.12 μm, *n* = 39 images), in time bin one, or the ventral region of the optic tract. The mean width for growth cone two changed to 16.3 μm (*SD* = 5.79 μm, *n* = 39 images), in the dorsal optic tract. For the two growth cones, we calculated an average decrease of width of approximately 4% (*SD* = 14.3%, *n* = two growth cones) as they navigated through the optic tract of the living brain preparation. This corresponded to an average rate of decrease in growth cone width of 0.6%/h (*SD* = 2.38%/h, *n* = two growth cones) during optic axon navigation from the ventral to the dorsal optic tract.

3.1.2. *Growth cone perimeter increases, while area decreases over time*

The perimeter and area of the growth cones were measured using the techniques described in Section 2. Measurements were taken for both growth cones in each of the 119 frames. Again, the time-lapse video was decomposed into three equal time bins and average growth cone perimeter and area were calculated for each of the time bins and plotted as bar graphs with trend lines (**Figure 4**). The results revealed that, on average, the growth cone perimeter increased, and the area decreased as the optic axons progressed through the optic tract (**Figure 5**). In addition, the growth cone perimeter increased more than the growth cone area decreased as the optic axons navigated through the optic tract in a living brain preparation (**Figure 5**).

The mean perimeter of growth cone one for time bin one was approximately 112 μm (*SD* = 13.5 μm, *n* = 39 frames), whereas the mean perimeter for growth cone one for the third time bin, corresponding to the dorsal optic tract, increased to 134 μm (*SD* = 30.7 μm, *n* = 39 images). The mean perimeter of growth cone two for time bin one was 98 μm (*SD* = 14.3 μm, *n* = 39 images), whereas the mean perimeter for growth cone two for the third time bin, corresponding to the dorsal optic tract, was 124 μm (*SD* = 26.3 μm, *n* = 39 images). This corresponds to an average increase of 23% (*SD* = 4.9%, *n* = two growth cones) for the perimeter of the growth cones as they navigated through the optic tract. We also calculated an average rate of increase in growth cone perimeter of 3.8%/h (*SD* = 0.8%/h, *n* = two growth cones) during their navigation through the optic tract.

The mean area of growth cone one for time bin one was 392 μm^2 (*SD* = 48.9 μm^2, *n* = 39 images), whereas the mean area for the third time bin, corresponding to the dorsal optic tract was decreased to 346 μm^2 (*SD* = 112.3 μm^2, *n* = 39 images). The mean area of growth cone two for time bin one was 333 μm² (*SD* = 71.1 μm², *n* = 39 images), whereas the mean area for growth cone two for the third time bin, corresponding to the dorsal optic tract was 321 μm² (*SD* = 71.3 μm², *n* = 39 images). On average, the two growth cones decreased their area by 8.6% (*SD* = 6.65%, *n* = two growth cones) during their navigation through the optic tract. This corresponds to a rate of decrease of area of 1.4%/h (*SD* = 1/1%/h, *n* = two growth cones) for optic axonal growth cones in the optic tract.

3.1.3. Filopodial protrusions increase, and lamellipodial protrusions decrease over time

To further decompose growth cone motility in the optic tract, the number of filopodial and lamellipodial protrusions in the growth cones was measured using criteria described in Section 2. The number of protrusions was calculated for both growth cones in each of the 119 frames of the time-lapse sequence. The average number of filopodia and lamellipodia for the growth cones for each of the time bins were calculated and plotted as bar graphs with trend lines (**Figure 6**). The results revealed that the mean number of filopodial protrusions increased, and the mean number of lamellipodial protrusions decreased, as the growth cones navigated through the optic tract toward the optic tectum (**Figure 6**). However, the mean number of filopodia per growth cone increased much more than the mean number of lamellipodia decreased during the time the optic axons extended through the optic tract (**Figure 6**).

The mean number of filopodia in growth cone one during time bin one (117 min) or the ventral optic tract was 4.4 (*SD* = 1.69, *n* = 39 images), whereas the mean number of filopodia for this growth cone for the third time bin, corresponding to the dorsal optic tract, was 10.3 (*SD* = 1.49, *n* = 39 frames). The mean number of filopodia displayed by growth cone two during time bin one was 3.8 (*SD* = 1.48, *n* = 39 images), whereas the mean number of filopodia for this growth cone in the third time bin, corresponding to the dorsal optic tract, increased to 10 (*SD* = 2.1, *n* = 39 images). On average, the two growth cones increased their average number of filopodia by 148% (*SD* = 16.6%, *n* = two growth cones), or at a rate of 25%/h (*SD* = 2.8%/h, *n* = two growth cones).

The mean number of lamellipodia of growth cone one for time bin one was 2.4 (*SD* = 0.97, *n* = 39 images), whereas the mean number of lamellipodia for the third time bin, corresponding to the dorsal optic tract, was 2.1 (*SD* = 0.68, *n* = 119 images). The mean number of lamellipodia of growth cone two for time bin one was 2.4 (*SD* = 0.75, *n* = 39 images), whereas the mean number of lamellipodia for the third time bin, corresponding to the dorsal optic tract decreased to 1.9 (*SD* = 0.63, *n* = 39 images). On average, the two growth cones decreased their average number of lamellipodia by 17% (*SD* = 5.3%, *n* = two growth cones), which corresponds to a mean rate of decrease of lamellipodia of 2.8%/h (*SD* = 0.87%/h, *n* = two growth cones).

4. Discussion

4.1. Changes in optic axonal growth cone morphologies in the optic tract

This quantitative analysis of growth cone motility of the *in vivo* time-lapse video demonstrates six morphological changes in growth cones of optic axons navigating in the optic tract of *X. laevis* brains. There was an increase of length, decrease of width, increase in perimeter, decrease in area, increase in number of filopodia, and a decrease in number of lamellipodia of the growth cones (**Figures 4–6**). These results show that as the growth cones progress through the optic tract, they become less circular, and more elongated and protrusive (**Figure 2**). These findings could help us gain a better understanding of the intracellular changes that may be influencing the morphology of the growth cones. From *in vitro* and *in vivo* studies, we know that growth cones comprise three domains: the C-domain composed of microtubules, the T-domain containing actin arcs, and the P-domain composed of actin as well (**Figure 1**). The P-domain has also filopodial protrusions, composed of actin bundles, and lamellipodial protrusions, composed of a meshwork of actin (**Figure 1**; [1–3]). The change in the morphology of the growth cone as it progresses through the optic tract could be due to the intracellular changes in these filaments. As the growth cone loses its circularity and becomes a more protrusive structure, microtubules could be consolidating in the axon shaft, and actin could be remodeling to influence the growth cone morphology once it reaches the dorsal optic tract. The changes in morphology of the growth cone could also be due to the different extracellular cues, such as Netrins, Wnts, Cadherins and CAMS, the growth cone encounters as it progresses through the optic tract [13]. One possibility is that there could be more and/or different extracellular cues in the dorsal tectum, causing the growth cone to project more filopodia. Since filopodia sense extracellular cues and are composed of actin bundles, this could mean that actin bundles take a prominent role, whereas microtubules in the C-domain of the growth cone begin to play a less important role in the structure of the growth cone. The predominance of bundled actin could also be attributed to decreased retrograde actin flow or to changes in motor driven transport proteins like Myosin II [14]. How actin and microtubules communicate these extracellular cues to intracellular changes could in turn be due to signaling molecules like APC that bind both actin and microtubules. Microtubules may become less predominant as the axons reach the end of the tract because of increased levels of APC within the growth cone [15–17]. APC could also influence retrograde actin flow in the growth cone. These intracellular molecular mechanisms underlying growth cone form and motility could be examined with further research. It is clear through analysis of the time-lapse video that specific morphological changes in the growth cone do occur in the optic tract. Our quantitative analysis can help us refine our understanding of the complicated intracellular mechanisms present within the growth cone.

4.2. Previous quantitative analysis of growth cones of optic axons *in situ*

Previous study quantified morphologies of growth cone of optic axons in the optic tract of brains from *X. laevis* tadpoles [18]. In this earlier study, growth cones of optic neurons injected with Lucifer yellow dye were examined in the optic tract in transverse sections of fixed brains. From the images of these growth cones, the author measured size parameters (area, length, and width), as well as number of processes in different regions of the optic tract. However, there were several differences between these previous measurements of growth cone parameters in fixed brains and the measurements we present here based on a living brain preparation. First, the previous study defined the base of the growth cone as the region of the axon where there was an abrupt thickening, or as the point halfway between the thickest region of the growth cone and the axon [18]. In contrast, we defined the base of the growth cone as the point where the first protrusion of the growth cone appeared on the axon [9]. In addition, in contrast to our measurement of width of the growth cone at its maximum point, the previous study measured growth cone width at 4–6 different points along the growth cone and averaged these measurements to obtain a final width [18]. Finally, and most significantly, our growth cone size measurements included the protrusions of the growth cone, whereas these previous growth cone dimensional parameters did not include filopodial or lamellipodial protrusions [18]. These differences in experimental approach (living versus fixed brains) and morphometric criteria (definition of base of growth cone, of width of growth cones, and inclusion or lack thereof of filopodia in growth cone size measurements) between the earlier study and our report clarify why our measured values for width, length, and area of growth cones of optic axons are significantly greater (~2× greater for length, width, and area) than those presented in the previous study. The previous study also applied different criteria for classifying filopodia and lamellipodia in optic axonal growth cones than we did in this study. The author considered filopodia to be protrusions (between 0.2 and 0.5 μm in width) that projected 2 μm or more from the growth cone surface, whereas lamellipodia were processes shorter than 2 μm [18]. In contrast, based on a different study, we considered any protrusion extending from the growth cone body between 1 and 5 μm in width as lamellipodia, and any protrusions less than 1 μm in width as filopodia [12]. However, despite these differences in classification of filopodia and lamellipodia, we counted similar numbers of protrusions in optic axonal growth cones in the ventral and mid-optic tract as in the earlier study.

4.3. Limitations and future directions for measurements of growth cones *in vivo*

In this quantitative analysis of growth cone morphology, researchers measured dimensions and protrusions of growth cones of optic axons manually outlining and delimiting boundaries of growth cones themselves (based on set criteria). One concern with having human researchers perform morphometric analyses is the potential variability in their delimitation of growth cone boundaries, and accordingly, the lack of reproducibility in their measurements. To circumvent this issue, we had five different researchers who make the same morphometric measurements on the two growth cones in each frame of the time-lapse sequence. We then averaged the values obtained by the different researchers to calculate our final values for size measurements of growth cone morphologies. Another approach to ensure reproducibility in quantitative analysis of growth cone morphology would be to have an automated computer program performing the measurements. However, before applying an automated approach, several issues would need to be resolved. First, the growth cones would need to be resolved with computer vision in a three-dimensional brain (**Figure 3**). Most automated algorithms work well on growth cones imaged *in vitro* (in two-dimensional cultures) but have difficulty establishing realistic boundaries for growth cones imaged in intact, three-dimensional tissues. Second, one would need to algorithmically define the base of the growth cone using a specific criterion, such as the point of the first protrusion on the axon near the growth cone. Currently, any automated computer program that is able to perform this type of analysis on extending axons and growth cones *in vivo* is not known.

In addition, in this study, we measured morphometric parameters for a relatively small number of growth cone of optic axons based on a time-lapse video captured of two fluorescently-labeled growth cones navigating in the optic tract of a single living brain preparation. Therefore, it is possible that our measurements are not representative of growth cones of optic axons in living brains generally. Instead, our growth cone measurements may be biased by the experimental conditions of this brain preparation. For example, the pressure exerted by the cover slip on the living brain preparation can alter the morphology of the growth cones as they navigate through the optic tract. To expand and generalize these morphometric measurements, we would need to make measurements on additional GFP expressing growth cones in different living brain preparations. An appropriate sample number would be 10–15 growth cones in five different living brain preparations. This would allow us to determine whether our morphometric measurements are generally representative of growth cones of optic axons from *X. laevis* tadpoles.

Limitations notwithstanding, the detailed measurements that we made of growth cone parameters advance our understanding of the dynamics of optic axon pathfinding in the optic tract of *X. laevis* brains *in vivo*. Our study establishes a template for the types of morphometric measurements that could be made from additional time-lapse video sequences of optic axons navigating in the optic tract of living brains. In the future, the motility parameters we measured for optic axonal growth cones of *Xenopus* brains could be compared to similar parameters obtained for growth cones of other types of neurons in different tissues. This would allow researchers to determine in a precise, quantitative manner how growth cone motility varies in different cell types and/or species. Moreover, this detailed quantitative analysis of growth cone motility of wild type optic axons will be fundamental for future studies examining how mechanical and molecular cues regulate the growth cone motility of optic axons *in vivo*. Finally, researchers aiming to develop computational visualizations of growth cone motility could use the quantitative parameters we measured for optic axonal growth cones to develop more accurate *in silico* representations of developing axonal connectivity in neuronal projections [19].

Acknowledgements

We thank Sonia Witte for making the time-lapse video of GFP-labeled optic axons in a living brain preparation from a *X. laevis* tadpole (http://www.pdn.cam.ac.uk/directory/christine-holt; also see Ref. [10]). We are grateful to Christine Holt for allowing us to use this time-lapse video as the basis for this morphometric study of optic axonal growth cones in the optic tract *in vivo*.

Funding for this project was provided by the Masters in Medical Health Sciences program in the College of Osteopathic Medicine at Touro University California. Publication made possible in part by support from the Berkeley Research Impact Initative (BRII) sponsored by the UC Berkeley Library.

Author details

Anokh Sohal[1], James Ha[1], Manuel Zhu[1], Fayha Lakhani[1], Kavitha Thiagaragan[2], Lauren Olzewski[1], Raagav Monakrishnan[1] and Tamira Elul[1]*

*Address all correspondence to: tamira.elul@tu.edu

1 Touro University California, Vallejo, California, United States of America

2 University of California, Berkeley, Berkeley, California, United States of America

References

[1] Vitriol EA, Zheng JQ. Growth cone travel in space and time: The cellular ensemble of cytoskeleton, adhesion and membrane. Neuron. 2012;**73**(6):1068-1081. DOI: 10.1016/j.neuron.2012.03.005

[2] Dent EW, Gertler FB. Cytoskeletal dynamics and transport in growth cone motility and axon guidance. Neuron. 2003;**40**(2):209-227

[3] Lowery LA, Van Vactor D. The trip of the tip: Understanding the growth cone machinery. Nature Reviews Molecular Cell Biology. 2009;**10**(5):332-343. DOI: 10.1038/nrm2679

[4] Mueller BK. Growth cone guidance: First steps towards a deeper understanding. Annual Review of Neuroscience. 1999;**22**:351-388. DOI: 10.1146/annurev.neuro.22.1.351

[5] Geraldo S, Gordon-Weeks PR. Cytoskeletal dynamics in growth cone steering. Journal of Cell Science. 2009;**122**(20):3595-3604. DOI: 10.1242/jcs.042309.

[6] Wiley A, Edalat K, Chiang P, Mora M, Mirro K, Lee M, Murh H, Elul T. GSK-3beta and alpha-catenin binding regions of beta-catenin exert opposing effects on the terminal ventral optic axonal projection. Developmental Dynamics. 2008;**237**(5):1434-1441. DOI: 10.1002/dvdy.21549

[7] Elul TM, Kimes NE, Kohwi M, Reichardt LF. N- and C-terminal domains of beta-catenin, respectively, are required to initiate and shape axon arbors of retinal ganglion cells *in vivo*. Journal of Neuroscience. 2003;**23**(16):6567-6575.

[8] Ettienne-Manneville S. Actin and microtubules in cell motility: Which one is in control? Traffic. 2004;**7**:470-477. DOI: 10.1111/j.1600-0854.2004.00196.x.

[9] Bray D, Chapman K. Analysis of micro spike movements on the neuronal growth cone. Journal of Neuroscience. 1985;**12**:3204-3213

[10] Harris WA, Holt CE, Bohoeffer F. Retinal axons with and without their somata, growing to and arborizing in the tectum of Xenopus embryos: A timelapse video study of single fibres *in vivo*. Development. 1987;**101**(1): 123-133

[11] Nieuwkoop PD, Faber J. Normal Table of *Xenopus laevis*. Amsterdam: Daudin; 1956

[12] Small JV, Stradal T, Vignal E, Rottner K. The lamellipodium: Where motility begins. Trends Cell Biology. 2002;**12**(3):112-120

[13] Hynes RO, Lander AD. Contact and adhesive specificities in the associations, migrations, and targeting of cells and axons. Cell. 1992;**68**(2):303-322

[14] Vallee RB, Seale GE, Tsai JW. Emerging roles for myosin II and cytoplasmic dynein in migrating neurons and growth cones. Trends Cell Biology. 2009;**7**:347-355. DOI: 10.1016/j.tcb.2009.03.009

[15] Purro SA, Ciani L, Hoyos-Flight M, Stamatakou E, Siomou E, Salinas PC. Wnt regulates axon behavior through changes in microtubule growth directionality: A new role for adenomatous polyposis coli. Journal of Neuroscience. 2008;**28**(34):8644-8654. DOI: 10.1523/JNEUROSCI.2320-08.2008

[16] Votin V, Nelson WJ, BArth AI. Neurite outgrowth involves adenomatous polyposis coli protein and beta-catenin. Journal of Cell Science. 2005;**118**(24):5699-5708. DOI: 10.1242/jcs.02679

[17] Zhou FQ, Zhou J, Dedhar S, Wu YH, Snider WD. NGF-induced axon growth is mediated by localized inactivation of GSK-3beta and functions of the microtubule plus end binding protein APC. Neuron. 2004;**42**(6):897-912. DOI: 10.1016/j.neuron.2004.05.011

[18] Holt CE. A single-cell analysis of early retinal ganglion cell differentiation in Xenopus: From soma to axon tip. Journal of Neuroscience. 1989;**9**:3123-3145

[19] Buettner HM. Computer simulation of nerve growth cone filopodial dynamics for visualization and analysis. Cell Motility and the Cytoskeleton. 1995;**32**(3):187-204. DOI: 10.1002/cm.970320304

Chapter 4

MRI Morphometry of the Brain and Neurological Diseases

Sergey Kotov

Additional information is available at the end of the chapter

http://dx.doi.org/10.5772/intechopen.69098

Abstract

The diagnosis of diseases of the brain is based on additional methods, confirming the clinical diagnosis. One of the most objective methods is magnetic resonance imaging (MRI). A detailed quantitative evaluation became possible after the introduction of MRI voxel-morphometry–statistical analysis of structural MRI images using a computerized segmentation matter of the brain gray and white matter. The decrease in the volume of the brain, as a manifestation of cerebral atrophy, is a common feature of many neurological diseases. We performed a study of brain structures in multiple sclerosis, Parkinson's disease, and cerebrovascular diseases. In patients with multiple sclerosis the correlation was found between the score on a scale of Expanded Disability Status Scale and the total thickness of the cerebral cortex. In our study of the brain in Parkinson's disease, the amount of the substantia nigra was slightly lower than in the control. In patients with long-following Parkinson's disease, the volume of substantia nigra was significantly higher than in patients with early stage. The increased volume was determined by the accumulation of organic iron compounds as a sign of neurodegeneration.

Keywords: magnetic resonance imaging, voxel morphometry, brain, gray matter, white matter, neurodegeneration, multiple sclerosis, Parkinson's disease

1. Introduction

The human brain is the most important organ that controls the functions of all the other organs in the human body through neuronal connectivity and neuronal signal transmission. Central nervous system (CNS) is the most complex but a very poorly understood structure in the human body. The diagnosis of many diseases of the brain is based on additional research methods, confirming the clinical diagnosis. One of the most objective and intuitive methods is magnetic resonance imaging (MRI). The technology to image the structure and function of the brain noninvasively (MRI, other methods of neuroimaging) has transformed our understanding of neurological disorders, opening new approaches to treatment and prevention. For a long time, MRI was based on a visual qualitative analysis of the obtained images. The development of the method led to the emergence of different modes of research, there are fundamentally new methods—diffusion-weighted image, the tensor image, perfusion MRI, functional MRI, and a method of MRI spectroscopy. However, these methods are mostly qualitative.

Neuroimaging may be divided into structural neuroimaging, which evaluates anatomic changes that occur in neurodegenerative conditions, and functional neuroimaging, which evaluates CNS activities such as blood flow and metabolism. For example, structural neuroimaging with computed tomography (CT) and magnetic resonance imaging (MRI) has defined patterns and rates of brain volume loss in neurodegeneration. Moreover, structural neuroimaging may detect treatable conditions that may present with dementia-like symptoms. Functional neuroimaging with MRI (fMRI), single photon emission CT (SPECT), and positron emission tomography (PET) have greatly assisted in the understanding of blood flow, metabolism, and amyloid deposition. Application of MR spectroscopy (MRS) has also yielded novel information about neurodegeneration [1, 2].

A detailed quantitative evaluation became possible after the introduction of MRI voxel-morphometry–statistical analysis of structural MRI images using a computerized segmentation matter of the brain GM and WM. Data about this technique were first published in 2000 [3]. MRI voxel morphometry requires the processing of data obtained by MRI study in the mode of the T1-weighted images (WI) with a slice thickness of 1 mm.

Voxel morphometry is based on the calculation of local differences in brain tissue after leveling expressed differences in anatomical structure and spatial position. This is achieved by spatial normalization (registration) of structural images into a single stereotactic space, further segmentation of gray and white matter, cortical rectification of furrows and convolutions, and statistical analysis to detect differences between the experimental groups. By segmentation, we can separate the major cerebral and extracerebral structures (GM, WM, and cerebrospinal fluid, CSF), also using voxel morphometry, we can reveal focal anatomic lesions [4, 5].

The decrease in the volume of the brain, as a manifestation of cerebral atrophy, is a common feature of many neurological diseases. There is the general (total) and regional (local) atrophy. Under the general atrophy, we can understand the volume reduction of brain parenchyma and the increase of subarachnoid spaces and ventricles of the brain. The decrease in the volume of certain brain structures is called regional atrophy. In brain tissue, the atrophy is accompanied by loss of neurons and connections between them that allows to consider it as one of the markers of the severity of the disease.

There are two basic methods of studying the volumes of the brain: using separation of gray and white matter manually or using special computer programs. Both methods have both positive and negative sides. We have tested both approaches. We performed a study of brain structures in multiple sclerosis, Parkinson's disease, and cerebrovascular diseases (CVD).

2. Morphometry of the brain in multiple sclerosis

Multiple sclerosis (MS) is a chronic inflammatory demyelinating disease of the CNS affecting more than 2 million people worldwide and leading to chronic progressive disability in the majority of cases. MS is heterogeneous both clinically and histopathologically, suggesting that different effector cells and molecular mechanisms are involved in the induction of tissue destruction. The most common form of MS, known as relapsing–remitting MS (RRMS), is associated with acute inflammatory episodes that reduce neurological function. RRMS patients may experience some recovery between relapses, but in 80% of cases, the disease evolves to a more progressive form with neurodegeneration termed secondary progressive MS (SPMS). The latter is associated with a gradual loss of neurological functions and is less dependent on inflammation [6–8]. Currently, it is generally assumed that immune, inflammatory, and neurodegenerative mechanisms involve in the pathogenesis of MS; it is a balance between the activity of inflammation, progressive degeneration, and reparative mechanisms and determines the clinical manifestations at each stage.

The main method of diagnosis in MS is magnetic resonance imaging (MRI). Standard techniques MRI is essential to confirm the nature determination of the activity of the pathological process and monitoring of the disease. Now, we use modern methods of MRI, which aims at clarifying the pathogenesis and pathophysiological mechanisms of formation of neurological deficit, the development of effective prognostic criteria, and new markers for the monitoring of the flow of inflammatory and neurodegenerative, atrophic component [9–11].

Initially, the presence of atrophy of the brain in MS patients was identified qualitatively, describing the extension of the ventricles and subarachnoid spaces and reducing the amount of substance of a brain. The next step was the quantitative assessment of atrophy of the brain with no differentiation of GM and WM. Currently, studies use different methods of MRI to assess how the global (brain in general) and regional atrophy. One of the most commonly used methods is to estimate the fraction of the brain parenchyma (BPF), which is determined by the ratio of the brain to the amount of the substance of the brain and cerebrospinal fluid (CSF). It is believed that in patients with MS, the regional atrophy may serve as a more sensitive marker of the severity of the disease than the general atrophy, but the results obtained in different studies often directly contradict each other. In recent years, it has been suggested that atrophy of cortical and subcortical gray matter prevails over the decrease in the volume of the white matter of the brain and largely determines the degree of disability of MS patients. Many researchers have noted that different types of MS course are characterized by their patterns of atrophy [12].

First, when an MS that the key importance was given to inflammatory demyelination of the WM and therefore considered that it undergoes atrophy. However, damage to axons within the WM lesions lead to atrophy, which was most likely evolving in two ways: the loss of substance in the demyelination and further Wallerian degeneration pathways associated with the lesion. WM atrophy affected certain areas of the brain, including both hemispheres of the brain, the brain stem, and the cerebellum. In patients with relapsing-remitting multiple sclerosis (RRMS) compared with the control group, most pronounced atrophy of almost all parts of the WM: the corpus callosum, cingulate gyri on both sides, certain divisions of the frontal lobes (including the upper sections of the radiate crown and the upper longitudinal beams), and temporal and occipital lobes (arch, upper and lower longitudinal beams, inferior frontal-occipital tracts [13, 14].

Atrophy of the brain in MS is the result of a comprehensive process of demyelination, axonal degeneration, and neuronal death [15]. In the treatment of MS, special attention is paid to the prevention of neurodegenerative component, and recently, several studies have been conducted for the evaluation of drugs. It was discovered that the degree of disability of the patients is determined with a therapeutic effect on a neurodegenerative component of the disease. However, the relationship is ambiguous. After the usage of MS-modifying drugs, there is a decrease in the degree of atrophy of the brain in the first year, then the physician can see the stabilization of process [16]. Thus, it may be the gradual, progressive atrophy of brain substance in MS but may be the fluctuations of its volume. In particular, acute inflammation with swelling of the brain tissue and the formation of new lesions during exacerbation of MS leads to a temporary increase in the volume of the brain and vice versa—glucocorticoid treatment leads to a temporary reduction of brain tissue—the so-called pseudoatrophy [17]. As suggested, this is the result of reducing inflammation and swelling in the brain.

Some studies have shown that WM atrophy is less prominent compared to GM atrophy due to more pronounced inflammatory processes that can mask atrophy [18, 19]. It is believed that patients with MS regional atrophy may serve as a more sensitive marker of the severity of the disease than the general, but the results obtained in different studies are mixed and often directly contradict each other. In recent years, it has been suggested that atrophy of cortical and subcortical GM prevails over the decrease in the volume of WM in the brain. It predominantly determines the degree of disability in patients with MS. However, many researchers note that different types of MS have the specific patterns of atrophy.

When analyzing the state of the GM in MS, it has been discovered that atrophy of the thalamus develops earlier than the atrophy of the cortex, which was demonstrated in a study of patients with RRMS and secondary progressive multiple sclerosis (SPMS) during follow-up [20, 21]. Atrophy of the thalamus observed in all types of MS course, to the greatest extent in SPMS, which is probably related to disease duration

The correlation was found between the score on a scale of Expanded Disability Status Scale (EDSS) and the total thickness of the cerebral cortex, precentral cortex, postcentral, parahippocampal, occipital gyri, and the caudate nucleus volume and a striped body. Some studies show that patients with sustained progression of disability by EDSS have a much higher rate of atrophic processes in comparison with patients with stable neurological symptoms. It is also believed that the amount of GM compared with WM volume is a more sensitive predictor of disability, measured by EDSS [22].

Cognitive impairment, including loss of memory, attention, and speed of information reproduction, has been reported in 70% of patients with MS, and they occur already at early stages of the disease (within the first 3 years). Patients with RRMS with cognitive impairment compared with patients without such revealed a decrease in volume of the brain in general and

GM of the cerebral cortex. Indeed, cortical atrophy is predictive of cognitive impairments because even mild cognitive changes are associated with significant thinning of the cerebral cortex. Also, a significant correlation was discovered with atrophy of the thalamus [23].

Thus, the modern MRI techniques, including MRI voxel-morphometry, significantly expand our understanding of the pathogenesis of MS. Numerous studies show that in addition to the WM atrophy, the GM atrophy in MS is already found in the early stages of the disease, and it progresses faster in healthy people, being a significant MRI predictor of the development of disability. Studies that show atrophy in different areas of the brain at the different types of MS course bring new contributions to the understanding of the pathophysiological mechanisms of the degenerative process. Now, we only accumulate our own data on the morphometry of the brain in MS.

We studied 10 patients with RRMS, five women, and five men. The diagnosis was based on McDonald criteria, 2010 revision. The age of the patients was from 24 to 36 years (average age/here and further data are given in the format M ± m/30, 7 ± 1, 22 years. At the time of the study, the disease duration ranged from 1 to 24 years (mean age of 8.1 ± 2.25 years). During follow-up, only three patients had one to three exacerbations. Before therapy with drugs modifying the course of the disease, mild disability (EDSS ≤ 3.0) was in eight patients, moderate (EDSS 3.5–5.5 points)—in two, and no severe disability (EDSS ≥ 6.0 points) were observed. The average EDSS score before the start of the specific therapy was 2.5 ± 0, 25, during an exacerbation—3.5 ± 0.34. All patients had no therapy before inclusion in the study. Therapy with glatiramer acetate (in a standard dose of 20 mg subcutaneously daily) was prescribed to three patients, but in the future, one of them was transferred to natalizumab therapy (in a standard dose of 300 mg infusion monthly) in the cause of aggressive course of the disease (three exacerbations). Seven patients had therapy with interferon beta 1-b in a dose of 9.6-million IU subcutaneously through the day. The control group (CONTROL) consisted of 10 healthy individuals, five men and five women, mean age of 21.3 ± 1.9 years.

To assess the condition of patients, we used the standard neurological examination, Expanded Disability Status Scale (EDSS), Multiple Sclerosis Functional Composite (MSFC) which includes the quantitative tests that assess lower limb function/walking (Timed 25-Foot Walk) and upper limbs (9-Hole Peg Test), hearing test on addition of a sequence of digits PASAT-3 (Paced Auditory Serial Addition Test), and several subtests of the Wechsler Adult Intelligence Scale: subtest "Resemblance," subtest "Digit span", subtest "Coding", subtest "Kohs Block Design Test ", Beck's depression scale, and Beck's anxiety scale.

Brain MRI was performed on "Initial Achieva 3.0 T" (Philips Medical System Nederland BV) with a magnetic field of 3 T. The study protocol included the use of T1-WI with slice thickness 1 mm, the distance between the slices–0 mm, and after contrast enhancement. The MRI was performed with the processing of the sequences for morphometry of anatomical MRI using FreeSurfer 5.3 (an open source software suite for processing and analyzing human brain MRI images) for segmentation and visualization of brain structures.

The results of voxel-based morphometry using the anatomical MRI T1-WI of the brain show a significant decrease in the cortical and subcortical GM volumes and WM in comparison to

CONTROL (**Table 1**). The volume of cortical GM in MS patients was 13.9% lower, while subcortical GM was 21.8%. The decline in GM of the thalamus was the highest—25.6%. The volume of WM was lower than in the CONTROL by 20.4%, and the greatest difference was found in relation to the corpus callosum—46.6%. At the same time, it was discovered an increase in the volume of the third ventricle by 31.9% and the lateral ventricles by 84.4%. It was indicated that the severity of reduction in volumetric parameters of the brain increased in proportion to the duration of MS.

It was not found a clear relationship between the level of reduction in GM and WM volume of the brain and EDSS. But there was a positive relationship between the amount of Hypo lesions in the white matter of the brain and EDSS. We discovered that the parameters of neuropsychological testing in patients with MS was worse than in CONTROL, and at exacerbation period, there was marked a more significant decline. We identified a probable correlation between reduced volumetric and neuropsychological parameters.

Parameters (mm^3)	CONTROL	MS
The total volume of cortical GM	521763.46	449010.5
The volume of subcortical GM1v	63915.5	50008.38
The total GM volume	696492.4	607442.001
The total volume of WM	503950.6	401106.7975
The volume of the lateral ventricles	14853.5	27396.225
Volume III ventricle	1221.7	1610.95
Volume IV ventricle	1764.6	1970.5
The volume of the cerebellum	158134.7	135249.45
The volume of the thalami	16330.5	12157.15
The hippocampal volume	10146.9	7818.45
The volume of the corpus callosum	3261.9	1740.55
The volume of hypointensive lesions in brain WM	899	8915.1
The volume of hypointensive lesions in brain GM	9.6	33.75

Table 1. The volume of the brain by MRI in patients with MS and CONTROL obtained by voxel morphometry (own data).

3. Morphometry of the brain in Parkinson's disease

Parkinson's disease (PD) is currently the most common among neurodegenerative diseases, its incidence is 200–300 cases per 100 thousand people. In Russia, according to epidemiological studies in some regions, the prevalence of PD is only 40–140 cases per 100 thousand population, which is considerably lower than in Western Europe and North America. The pathogenesis of PD is of great importance to the accumulation of intracellular inclusions (Levi bodies) in the neurons located in the compact part of the substantia nigra (SN), reflecting progressive neurodegeneration. Motor symptoms of PD are manifested at the stage when it is killed not less than 50% of dopaminergic neurons of the SN, and the typical picture of the disease is fully formed during the destruction of 80% of the cells [24].

Due to the decrease in the number of neurons of the SN, the number of dopaminergic projections at the rear parts of the putamen is reduced that leads to the development of a triad of characteristic motor symptoms of PD—static tremor, bradykinesia, and rigidity, which is deployed in the next stages of the disease with postural instability. In the early stages of PD, not all symptoms are expressed equally; therefore, there are difficulties of differential diagnostics of PD and symptomatic parkinsonism, torsion dystonia, and essential tremor that are not associated with degeneration of nigral neurons and a deficit of dopamine in the striatum, while in 15% of cases, the diagnosis of PD is wrong [25–28]. The complexity of differential diagnosis and, to some extent, delayed the debut of the classic clinical symptoms leading to late diagnosis of the neurodegenerative process and the low efficiency of symptomatic treatment, while the possibilities for pathogenetic help have been lost.

At this time, the researchers are studying actively the predictive value of various non-motor symptoms that occur in the early stages of PD, as these symptoms of neurodegeneration manifest long before the development of the classic triad of motor symptoms. Known early signs of violations of smell, changes in the regulation of cardiac activity, disorders of motility of the gastrointestinal tract, ultrasound signs of early degeneration, disaster, and other changes, because they precede the manifestation of motor symptoms of PD in the years.

As a result of improving traditional methods of functional neuroimaging, the studies have been aimed at identifying qualitative and quantitative indicators of the morphology of the brain in PD. In this connection, the special interest is the study of dopaminergic structures of the brain stem. Despite the fact that currently there are data from multiple studies, obtained facts are rather contradictory. The difference in the results of volumetric studies of the SN in PD could be attributed to the difficulty of identifying the boundaries between the compact and reticular parts of the SN, lack of opportunity of contrasting these structures, individual differences in the volume of the midbrain and medulla in general [28, 29].

Using diffusion tensor MRI Tessa et al. [30] revealed the signs of a diffuse reduction in the volume of gray matter of the cerebral hemispheres in patients with newly diagnosed PD who did not receive specific therapy. These data are consistent with the hypothesis of PD stages of Braak [25], claiming that at the time of the manifestation of motor symptoms process of neurodegeneration is already becoming common. The authors found differences in the severity of atrophy with a predominance of rigidity or tremor obtained by using sophisticated statistical techniques, however, do not seem convincing. Some researchers using this technique revealed changes in the thalamus, SN, and its ascending and descending pathways in PD [31–33].

T1-, T2-, and T2*-WI were used previously to study the anatomy of the midbrain, it is assumed that T1-WI better reflects the lower divisions, however, none of the modes did not allow to distinguish the compact and reticular parts of CHS [34, 35]. Routine T1- and T2-WI MRI do not reveal accurate structural changes in the SN in PD, which confirms its idiopathic nature.

Morphometric studies, using voxel morphometry obtained by different researchers, are contradictory, probably due to the difficulty of tracing the boundaries of the compact part of SN.

The contrast ratio of the SN is determined by the accumulation in it neuromelanin, ferritin and other iron substances, whose concentration is changed by natural aging, so in PD and other neurodegenerations. Reticular part of SN contains more iron than its compact part, which is manifested by hypotensivity of this area due to the decrease of the relaxation time T2. The compact part of SN contains more neuromelanin, and the iron atoms in them are in the bound state with the ferritin [36, 37].

Oikawa et al. [38] did not note significant changes of size and density of CHS in studies in modes T2 and proton density. Minati et al. [39] using the T1 with the suppression of the signal from gray and white matter in PD revealed hypointensity signal from the external part of the SN, mainly the reticular part of it. The authors in the study of SN at the level of the chiasm above the upper legs of the cerebellum said that there is significantly smaller area in patients with PD (72.2 ± 27.4 mm^2) than in healthy individuals at similar (88.8 ± 28.7 mm^2) or young age (of 91.8 ± 29.4 per mm^2). The authors have not performed the volumetric study.

We carried out the clinical and neuroimaging examination of 4 groups of patients: 10 patients with the initial manifestations of PD, stage 1 on Hoehn and Yahr scale (PD1) [40], 10 patients with severe clinical manifestations, stage 3 on Hoehn and Yahr scale (PD2), 10 patients with early manifestations of CVD, after a transient ischemic attack (TIA) or minor/lacunar stroke with complete or almost complete regression of neurological deficits (CVD1), and 10 patients with severe manifestations of CVD after suffering repeated small/lacunar stroke (CVD2). The diagnosis of PD was established in accordance with generally accepted criteria of brain Bank of the UK PD society [41]. Control group (CONTROL) consisted of 10 apparently healthy persons aged 21–48 years.

MRI study was conducted on "Initial Achieva 3.0 T" (Philips Medical System Nederland BV) with a magnetic field of 3 T. The study protocol included the use of T2- and T2*-weighted images (T2-Wi, T2*-WI) using pulse sequences TSE (Turbo spin echo), FFE (Fast field echo), GraSE (Gradient spin echo). The orientation of slices was conducted nearly to the axial plane, parallel to the line connecting the middle of the chiasm and the lower contour of the corpus callosum splenium. To explore dopaminergic structures of the brain stem, the middle block of the scanned sections was located between the upper and lower hillocks of the corpora quadrigemina. Also, we carried out the scanning in the coronal and axial planes, perpendicular to the main axial plane.

Regional assessment of the volume of brain matter of cerebral hemispheres and the CSF, GM, and WM of the midbrain were performed by the volumetric morphometry. For the region of the cerebral hemispheres, we used axial slices extending above the cavity of the third ventricle, through the bodies of the lateral ventricles, above the insula. After visual determination of the zone of interest, we performed automatic segmentation on the GM + WM and CSF.

To determine the volumetric characteristics of SN, we investigated sections of the midbrain in T2-WI and FLAIR. T2 and T2* relaxation time changes as a result of the accumulation of iron in the area of SN in PD patients (**Figure 1**). The highest contrast of images we obtained in the FLAIR MRI. Study of SN included in its entirety as Hypo-region compared to the white matter

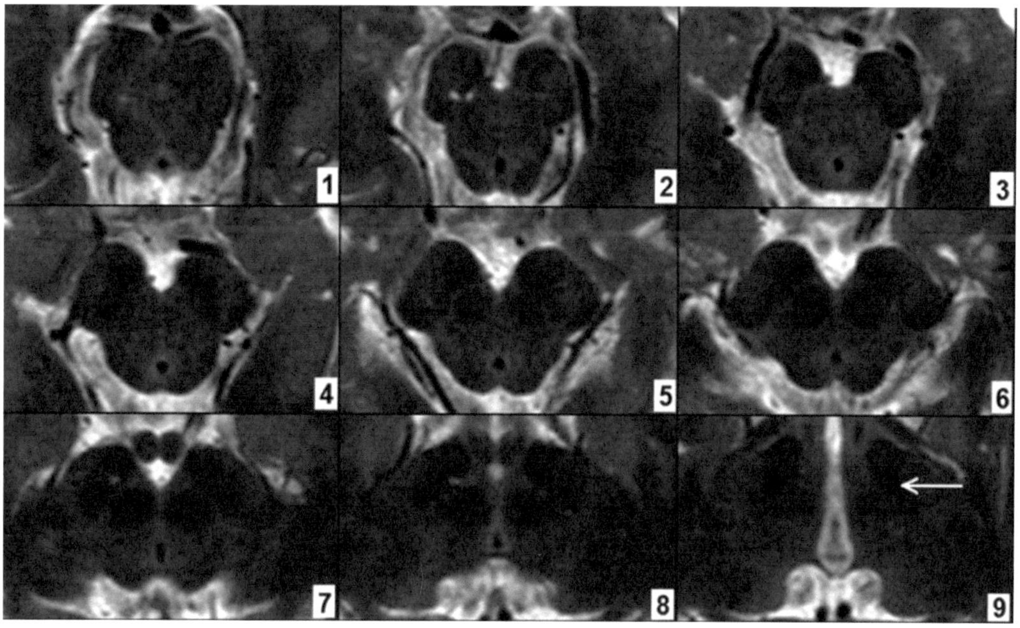

Figure 1. Images of the midbrain from the ponto-mesencephalic (1) to mesencephalic-diencephalic transition (9) on T2-WI (slice thickness 2 mm, the distance between the slices–0 mm). In Sections 3–8 visualized SN, in Sections 5–9—NR, slice 9–the subthalamic nucleus (arrow) positioned as a continuation of SN at the diencephalic level (own data).

of the midbrain. The normalization of the size SN on the volume of the midbrain did not perform. For accurate tracing of the boundaries of SN and the nucleus ruber (NR), we used schemes Duvernoy's Atlas of the Human Brain Stem and Cerebellum [42].

The boundaries of SN and NR were traced manually. The border was carried out by the regions, having approximately the intermediate brightness signal between SN and surrounding white matter. To reduce the subjective factor, the study was conducted by two experts independently from each other.

A significant decrease in the brain volume compared to the CONTROL was observed only in the groups CVD1 and CVD2, while the increase in the volume of subarachnoid spaces was observed in all patients. The PD1 group showed a trend toward increasing the share of space filled with CSF in the preservation compared to the normal shares of the brain tissue. Patients of the PD2 group, along with the increase in the share of space filled with CSF, showed a trend toward a decrease of the relative volume of brain matter. The greatest reduction in the volume of brain matter with a corresponding increase space filled with CSF was observed in the CVD2 group. The difference was statistically significant with the PD1 and PD2 groups (**Table 2**).

We carried out morphometry of SN in PD [43]. Although the amount of SN in PD1 patients was slightly lower than in the CONTROL, but the difference was not significant. At the same time in PD2 patients, the volume of SN was significantly higher than in PD1 patients (**Figure 2**). It should be emphasized that the indicators include both compact and reticular part of SN, and hypotensive of this zone was determined by the accumulation neuromelanin, ferritin, and

Group	Volume of brain matter	Volume of the lateral ventricles	Volume of the subarachnoid space
PD1	0.85 ± 0.01	0.04 ± 0.003 A	0.11 ± 0.01 A
PD2	0.82 ± 0.01 A	0.06 ± 0.01 B	0.12 ± 0.01 A
CVD1	0.81 ± 0.01 A, B	0.07 ± 0.004 B	0.12 ± 0.01 A
CVD2	0.78 ± 0.01 A, B, C	0.09 ± 0.01 A, B, C	0.12 ± 0.01 A
CONTROL	0.87 ± 0.01	0.06 ± 0.01	0.07 ± 0.01

Abbreviations: A, $p < 0.01$ value is calculated against CONTROL; B, $p < 0.01$ value is calculated against PD1; C, $p < 0.05$ value is calculated against PD2 by volume of the brain and $p < 0.01$ for the volume of the lateral ventricles.

Table 2. The ratio of the volumes of brain matter and spaces filled with CSF, in patients with PD, CVD, and control group (M ± m) (own data).

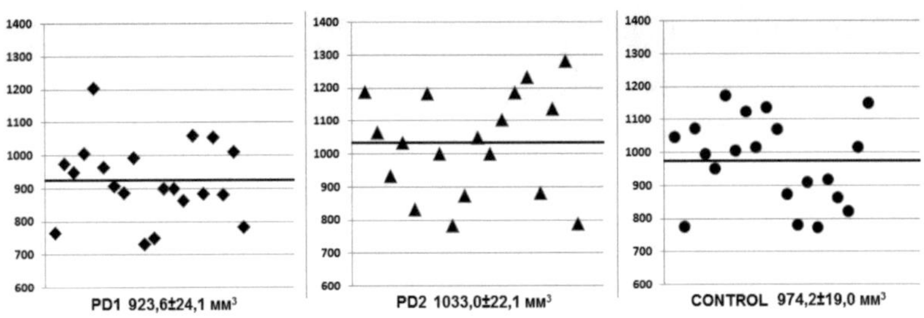

Figure 2. Volume of SN according to the results of morphometry axial slices (in mm³) in the T2-Wi and FLAIR in PD1, PD2, and CONTROL. The horizontal line indicates the median in each group (own data).

other organic iron compounds. Previously, researchers revealed hyperechogenicity of the SN in patients with PD in transcranial ultrasonography. Obviously, in this case, MRI revealed a phenomenon similar to ultrasonography.

Thus, MRI neuroimaging of dopaminergic structures of the brain stem is feasible, it can detect volumetric changes reflecting a neurodegenerative process. Although the positron emission tomography allows to obtain more information regarding the functional state of the dopaminergic structures in PD; however, this method remains inaccessible. Therefore, voxel morphometry of SN is a useful method of diagnosis and monitoring of PD patients.

Author details

Sergey Kotov

Address all correspondence to: kotovsv@yandex.ru

Neurologic Department of Moscow Regional Research and Clinical Institute n.a. M.F. Vladimirsky ("MONIKI"), Moscow, Russia

References

[1] Wippold FJ 2nd, Brown DC, Broderick DF, Burns J, Corey AS, Deshmukh TK, Douglas AC, Holloway K, Jagadeesan BD, Jurgens JS, Kennedy TA, Patel ND, Perlmutter JS, Rosenow JM, Slavin K, Subramaniam RM. ACR appropriateness criteria dementia and movement disorders. Journal of the American College of Radiology. 2015;**12**(1):19-28. DOI: 10.1016/j.jacr.2014.09.025

[2] Suckling J, Nestor LJ. The neurobiology of addiction: The perspective from magnetic resonance imaging present and future. Addiction. Feb 2017;**112**(2):360-369. DOI: 10.1111/add.13474

[3] Ashburner J, Friston KJ. Voxel-based morphometry—The methods. NeuroImage. 2000;**11**:805-821

[4] Wright IC, McGuire PK, Poline JB, Travere JM, Murray RM, Frith CD, Frackowiak RS, Friston KJ. A voxel-based method for the statistical analysis of gray and white matter density applied to schizophrenia. Neuroimage. 1995;**2**(4):244-252

[5] Wu P, Zhou YM, Zeng F, Li ZJ, Luo L, Li YX, Fan W, Qiu LH, Qin W, Chen L, Bai L, Nie J, Zhang S, Xiong Y, Bai Y, Yin CX, Liang FR. Regional brain structural abnormality in ischemic stroke patients: A voxel-based morphometry study. Neural Regeneration Research. 2016;**11**(9):1424-1430

[6] Compston A, Coles A. Multiple sclerosis. Lancet. 2008;**372** (9648):1502-1517. DOI: 10.1016/S0140-6736(08)61620-7

[7] Lucchinetti C, Brück W, Parisi J, Scheithauer B, Rodriguez M, Lassmann H. Heterogeneity of multiple sclerosis lesions: implications for the pathogenesis of demyelination. Annals of Neurology. 2000;**47**(6):707-717

[8] Milo R, Kahana E. Multiple sclerosis: Geoepidemiology, genetics and the environment. Autoimmunity Reviews. 2010;**9**(5):A387-94. DOI: 10.1016/j.autrev.2009.11.010

[9] Polman CH, Reingold SC, Banwell B, Clanet M, Cohen JA, Filippi M, Fujihara K, Havrdova E, Hutchinson M, Kappos L, Lublin FD, Montalban X, O'Connor P, Sandberg-Wollheim M, Thompson AJ, Waubant E, Weinshenker B, Wolinsky JS. Diagnostic criteria for multiple sclerosis: 2010 revisions to the McDonald criteria. Annals of Neurology. 2011;**69**(2):292-302. DOI: 10.1002/ana.22366

[10] Uher T, Horakova D, Kalincik T, Bergsland N, Tyblova M, Ramasamy DP, Seidl Z, Vaneckova M, Krasensky J, Havrdova E, Zivadinov R. Early magnetic resonance imaging predictors of clinical progression after 48 months in clinically isolated syndrome patients treated with intramuscular interferon β-1a. European Journal of **Neurology**. 2015;**22**(7):1113-1123. DOI: 10.1111/ene.12716

[11] Sormani MP, Arnold DL, De Stefano N. Treatment effect on brain atrophy correlates with treatment effect on disability in multiple sclerosis. Annals of Neurology. 2014;**75**(1):43-49. DOI: 10.1002/ana.24018

[12] Radue EW, Barkhof F, Kappos L, Sprenger T, Häring DA, de Vera A, von Rosenstiel P, Bright JR, Francis G, Cohen JA. Correlation between brain volume loss and clinical and MRI outcomes in multiple sclerosis. Neurology. 2015;**84**(8):784-793. DOI: 10.1212/WNL.0000000000001281

[13] Sailer M, Fischl B, Salat D, Tempelmann C, Schonfeld MA, Busa E, Bodammer N, Heinze HJ, Dale A. Focal thinning of the cerebral cortex in multiple sclerosis. Brain 2003;**126**:1734-1744

[14] Moccia M, Quarantelli M, Lanzillo R, Cocozza S, Carotenuto A, Carotenuto B, Alfano B, Prinster A, Triassi M, Nardone A, Palladino R, Brunetti A, Brescia Morra V. Grey: white matter ratio at diagnosis and the risk of 10-year multiple sclerosis progression. European Journal of Neurology. 2017;**24**(1):195-204. DOI: 10.1111/ene.13183

[15] Lansley J, Mataix-Cols D, Grau M, Radua J, Sastre-Garriga J. Localized grey matter atrophy in multiple sclerosis: A meta-analysis of voxel-based morphometry studies and associations with functional disability. Neuroscience & Biobehavioral Reviews. 2013;**37**(5):819-830. DOI: 10.1016/j.neubiorev.2013.03.006

[16] Jacobsen C, Hagemeier J, Myhr KM, Nyland H, Lode K, Bergsland N, Ramasamy DP, Dalaker TO, Larsen JP, Farbu E, Zivadinov R. Brain atrophy and disability progression in multiple sclerosis patients: A 10-year follow-up study. Journal of Neurology, Neurosurgery, and Psychiatry. 2014;**85**(10):1109-1115. DOI: 10.1136/jnnp-2013-306906

[17] Zivadinov R, Reder AT, Filippi M, Minagar A, Stüve O, Lassmann H, Racke MK, Dwyer MG, Frohman EM, Khan O. Mechanisms of action of disease-modifying agents and brain volume changes in multiple sclerosis. Neurology. 2008;**71**(2):136-144. DOI: 10.1212/01.wnl.0000316810.01120.05

[18] Carone DA, Benedict RH, Dwyer MG, Cookfair DL, Srinivasaraghavan B, Tjoa CW, et al. Semi-automatic brain region extraction (SABRE) reveals superior cortical and deep gray matter atrophy in MS. Neuroimage. 2005;**29**(2):505-514

[19] Morgen K, Sammer G, Courtney SM, Wolters T, Melchior H, Blecker CR, Oschmann P, Kaps M, Vaitl D. Evidence for a direct association between cortical atrophy and cognitive impairment in relapsing-remitting MS. NeuroImage. 2006;**30**:891-898

[20] Audoin B, Davies GR, Finisku L, Chard DT, Thompson AJ, Miller DH. Localization of grey matter atrophy in early RRMS: A longitudinal study. NeuroImage. 2006;**253**(11): 1495-1501

[21] Sepulcre J, Sastre-Garriga J, Cercignani M, Ingle GT, Miller DH, Thompson AJ. Regional gray matter atrophy in early primary progressive multiple sclerosis: A voxel-based morphometry study. Archives of Neurology. 2006;**63**(8):1175-1180

[22] Fisniku LK, Chard DT, Jackson JS, Anderson VM, Altmann DR, Miszkiel KA, Thompson AJ, Miller DH. Gray matter atrophy is related to longterm disability in multiple sclerosis. Annals of Neurology. 2008;**64**:247-254

[23] Calabrese M, Rinaldi F, Mattisi I, Grossi P, Favaretto A, Atzori M, Bernardi V, Barachino L, Romualdi C, Rinaldi L, Perini P, Gallo P. Widespread cortical thinning characterizes patients with MS with mild cognitive impairment. Neurology. 2010;**74**(4):321-328. DOI: 10.1212/WNL.0b013e3181cbcd03

[24] Braak H, Del Tredici K, Rüb U, de Vos RA, Jansen Steur EN, Braak E. Staging of brain pathology related to sporadic Parkinson's disease. Neurobiology of Aging. 2003;**24**(2): 197-211

[25] Braak H, Rub U, Jansen Steur EN, Del Tredici K, de Vos RA. Cognitive status correlates with neuropathologic stage in Parkinson disease. Neurology. 2005;**64**(8):1404-1410

[26] Whone AL, Watts RL, Stoessl J, et al. Slower progression of PD with ropinirole versus L-dopa: The REAL-PET study. Annals of Neurology. 2003;**54**(1):93-101

[27] Schneider SA, Obeso JA. Clinical and pathological features of Parkinson's disease. Current Topics in Behavioral Neurosciences. 2015;**22**:205-220. DOI: 10.1007/7854_2014_317

[28] Brooks DJ. Imaging approaches to Parkinson disease. Journal of Nuclear Medicine. 2010;**51**(4):596-609

[29] Geng DY, Li YX, Zee CS. Magnetic resonance imaging-based volumetric analysis of basal ganglia nuclei and substantia nigra in patients with Parkinson's disease. Neurosurgery. 2006;**58**(2):256-262

[30] Tessa C, Giannelli M, Della Nave R, Lucetti C, Berti C, Ginestroni A, Bonuccelli U, Mascalchi M. A whole-brain analysis in De Novo Parkinson disease. American Journal of Neuroradiology. 2008;**29**(4):674-680

[31] McKeown MJ, Uthama A, Abugharbieh R, Palmer S, Lewis M, Huang X. Shape (but not volume) changes in the thalami in Parkinson disease. BMC Neurology. 2008;8:8. DOI: 10.1186/1471-2377-8-8

[32] Menke RA, Scholz J, Miller KL, Deoni S, Jbabdi S, Matthews PM, Zarei M. MRI characteristics of the substantia nigra in Parkinson's disease: A combined quantitative T1 and DTI study. Neuroimage. 2009;**47**(2):435-441

[33] Vaillancourt DE, Spraker MB, Prodoehl J, Abraham I, Corcos DM, Zhou XJ, Comella CL, Little DM. High-resolution diffusion tensor imaging in the substantia nigra of de novo Parkinson disease. Neurology. 2009;**72**(16):1378-1384

[34] Manova ES, Habib CA, Boikov AS, Ayaz M, Khan A, Kirsch WM, Kido DK, Haacke EM. Characterizing the mesencephalon using susceptibility-weighted imaging. American Journal of Neuroradiology. 2009;**30**(3):569-574

[35] Sasaki M, Shibata E, Tohyama K, Takahashi J, Otsuka K, Tsuchiya K, Takahashi S, Ehara S, Terayama Y, Sakai A. Neuromelanin magnetic resonance imaging of locus ceruleus and substantia nigra in Parkinson's disease. Neuroreport. 2006;**17**(11):1215-1218

[36] Double K, Gerlach M, Schünemann V, Trautwein AX, Zecca L, Gallorini M, Youdim MB, Riederer P, Ben-Shachar D. Iron-binding characteristics of neuromelanin of the human substantia nigra. Biochemical Pharmacology. 2003;**66**(3):489-494

[37] Zecca L, Gallorini M, Schuenemann V, Trautwein AX, Gerlach M, Riederer P, Vezzoni P, Tampellini D. Iron, neuromelanin and ferritin content in the substantia nigra of normal subjects at different ages: Consequences for iron storage and neurodegenerative processes. Journal of Neurochemistry. 2001;**76**(6):1766-1773

[38] Oikawa H, Sasaki M, Tamakawa Y, Ehara S, Tohyama K. The substantia nigra in Parkinson disease: Proton density-weighted spin-echo and fast short inversion time inversion-recovery MR findings. American Journal of Neuroradiology. 2002;**23**(10):1747-1756

[39] Minati L, Grisoli M, Carella F, De Simone T, Bruzzone MG, Savoiardo M. Imaging degeneration of the substantia nigra in Parkinson disease with inversion-recovery MR imaging. American Journal of Neuroradiology. 2007;**28**(2):309-313

[40] Hoehn MM, Yahr MD. Parkinsonism: Onset, progression, and mortality. Neurology. 1967;**17**(5):427-442

[41] Hughes AJ, Daniel SE, Blankson S, Lees AJ. A clinicopathologic study of 100 cases of Parkinson's disease. Archives of Neurology.1993;**50**(2):140-148

[42] Naidich ThP, Duvernoy HM, Delman BN, Sorensen AG, Kollias SS, Haacke EM. Duvernoy's Atlas of the Human Brain Stem and Cerebellum. Wien: Springer-Verlag; 2009. pp. 53-116

[43] Bogdanov RR, Manannikova EI, Abrabenko AS, Maratkanova TV, Kotov SV. Morphometric parameters of the neurodegenerative process in Parkinson's disease and chronic cerebral ischemia. [Article in Russian] Zh Nevrol Psikhiatr Im S S Korsakova. 2013; **113**(10):40-44

Chapter 5

Application of Morphometric and Stereological Techniques on Analysis and Modelling of the Avian Lung

John N. Maina

Additional information is available at the end of the chapter

http://dx.doi.org/10.5772/intechopen.69062

> 'I often say that when you can measure what you are speaking about, and express it in numbers, you know something about it; but when you cannot express it in numbers, your knowledge is of a meager and unsatisfactory kind; it may be the beginning of knowledge, but you have scarcely, in your thoughts, advanced to the stage of science, whatever the matter may be'.
>
> *Lecture 'Electrical Units of Measurement' (3 May 1883)*
>
> *William Thompson (Lord Kelvin) (1824–1907).*

Abstract

For a long time, biology was a qualitative (descriptive) science. The investigations failed to fully explicate the functional designs of whole organisms and their constituent parts. About half a century ago, at an interdisciplinary meeting which was held in Feldberg (Germany), the International Society of Stereology (ISS) was formed. Mathematicians, statisticians and physical and biological scientists combined their skills to create a new scientific discipline of stereology that allowed for reliable and reproducible quantitation of structural entities of composite physical and biological materials and extrapolation of measurements made on two-dimensional profiles/images to their three-dimensional forms. With time, novel bias-free sampling and quantitation techniques have been developed and tested. Presently, there is no justification for totally descriptive biological studies. Numerous books, publications, computer programmes and applications and dedicated microscopes exist for cost-effective analysis. Within the relatively short time, it has been in existence, the ISS has actively advanced stereology which is now applied by scientists all over the world in various biological disciplines. Only basic understanding of mathematics, geometry and statistics is needed to do good stereology. Here, analysis of the avian (bird) lung is given to show the versatility and robustness of stereological techniques in analysing biological structures.

Keywords: stereology, morphometry, sampling, avian lung

1. Introduction

In biological studies, by default or design, the primary objective is to understand the relationships between the structure and the function of an organism or animal. The nature and organization of the cellular and tissue constituent parts specify its form and performance. For a long time, in rigidly compartmentalized scientific disciplines, it was not possible to correlate these aspects mainly because disciplines such as physiology and biochemistry generated functional (i.e. quantitative) data, structural studies such as morphology and anatomy were fundamentally descriptive. During the last half century, however, methods that allow biological structures to be meaningfully analysed have been developed by essentially synthesizing statistical and mathematical methods. Biology has profoundly changed from a mere observational, descriptive and classificatory discipline into a fully fledged branch of science with complex theories and concepts of quantitation and formulation of comprehensive mathematical models that powerfully explicate the forms and functions of complex dynamic systems. Quantifying and comparing different biological systems that share common design principles powerfully explain evolutionary and adaptive changes [1].

Quantitative studies require careful planning and execution. Before performing such a study, thorough understanding of the structure of a biological entity and its formative parts is imperative. A pilot study should precede the study and proper research questions formulated from it. A well-planned and properly executed quantitative study must adequately support or refute articulated research questions: *data should not be acquired for their own sake.*

The tissues analysed must be handled and processed carefully. This is to avoid introduction of artefacts during stages like sampling, processing, fixing and staining. The primary goal should be that the tissue analysed is preserved in its initial (natural) state. Since only a small fraction of the organ or tissue can routinely be analysed (**Figure 1**), rigorous unbiased and reproducible sampling is critical to meaningful quantitative studies [2, 3].

Morphometry means measurement of form (**morph** = form and **metric** = measurement), while stereology, which translates from its Greek roots 'stereo' and 'logos', means the 'science of studying solids'. Specifically, it entails extrapolation of measurements made on two-dimensional (2D) profiles/images to their three-dimensional (3D) configuration. In stereological studies, this entails reconstruction of the structural profiles to their in-life state. However, stereology is morphometry, but morphometry is not necessarily stereology.

A tailor who takes measurements of a person in order to make him or her a suit performs morphometry and not stereology. In the use of the highly informative transmission microscopes, be it light or electron, thin sections of tissues have to be cut to allow light or electron beam to pass through. The inescapable action converts 3D structures to 2D profiles (**Figure 2**). The profiles that are generated after sectioning cells, tissues or organs depend on the plane of sectioning (**Figure 3**). For example, circular profiles can be generated by sectioning tubular or spherical structures. *In order to correctly interpret such profiles, microscopists should think in 3D.* In biology, by applying stereological techniques, measurements made on 2D profiles can be extrapolated to their 3D configuration (**Figure 4**).

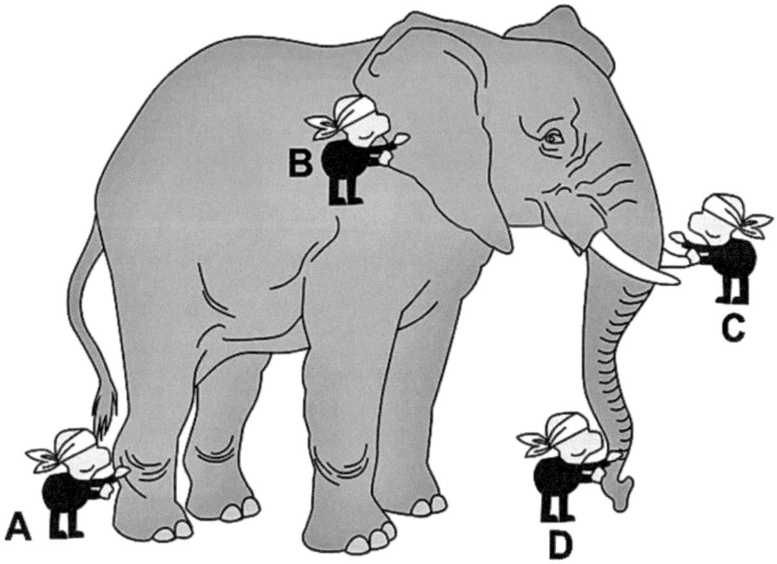

Figure 1. Cartoon showing three blind-folded people palpating different parts of an elephant. The individuals will describe the shape of the elephant differently. Person (A) will, e.g., describe it as corresponding to a trunk of a large tree trunk; (B) will describe it as smooth, flat object; (C) will pronounce it as hard and sharp-pointed; and (D) will deduce it as a soft, pliable object. In principle, all of them are correct. The figure illustrates the following: in quantitative studies, since a whole animal, organ or tissue cannot be analysed piece-by-piece, rigorous sampling has to be undertaken so as to acquire a representative and manageable sample for analysis.

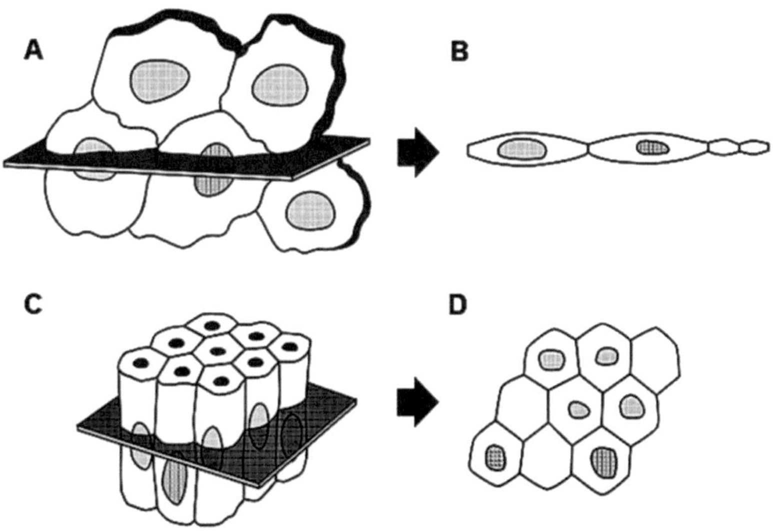

Figure 2. Diagrams showing transversely sectioned epithelial cells (A) and the profiles generated (B) and structures such as muscle fibres (C) that generate profiles when sectioned transversely (D). Stereology allows one to extrapolate measurements made on two-dimensional profiles (B and D) to their three-dimensional configuration.

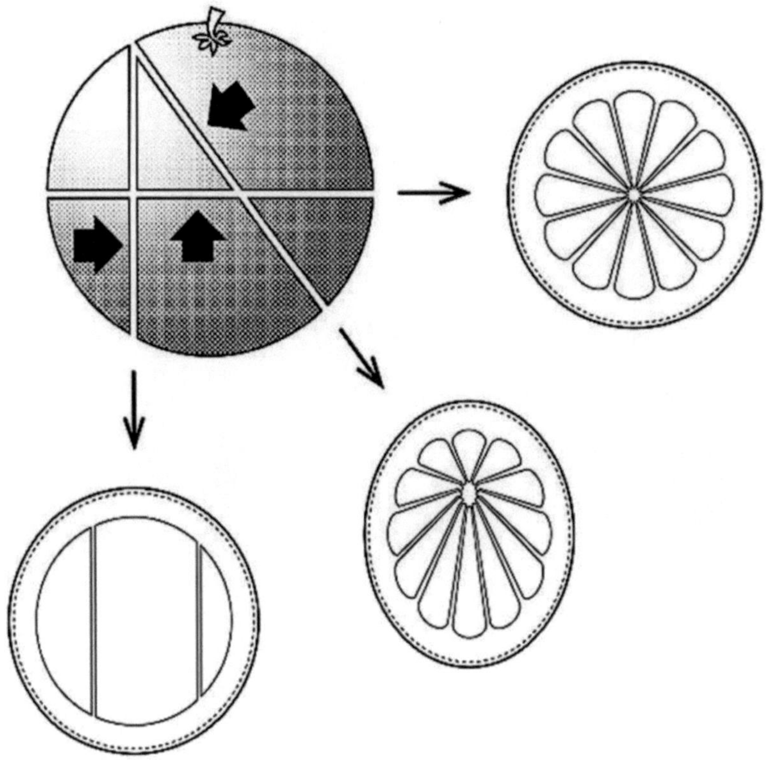

Figure 3. Diagram showing the different profiles that are generated after sectioning a fruit (e.g. an orange) at different planes. By applying stereological methods, the shape of the fruit can be reconstructed from studying sufficient random sections of the fruit.

As a scientific discipline, stereology did not formally start until ~55 years ago at a meeting of researchers from as varied disciplines as biology, geology, engineering and materials sciences in 1961. A biologist, Professor Hans Elias (specifically a histologist), organized a meeting at Feldberg in Germany which was attended by scientists with one goal in mind: *to quantify 3D images by studying their 2D sections (profiles).* Now called 'The International Society for Stereology and Image Analysis (ISS & IA)', the International Society for Stereology (ISS) was formed at the first congress in Vienna (Austria) in 1962 [4]. Since then, meetings are held regularly, and the ISS is a leading multi-disciplinary collaboration group.

The start of stereology in the 1960s occurred simultaneously with important technological developments that included availability of powerful and affordable microscopes and introduction of innovative tissue preparation and staining techniques such as immunocytochemistry. These developments increased interest in quantitative biology in general and in the emerging field of stereology in particular. Investigators began to prefer the more objective stereological approaches over subjective evaluations that were known to be greatly affected by inter-observer errors. Mathematicians and statisticians in particular have greatly contributed their unique skills to the theoretical aspects of the interdisciplinary discipline of stereology [5]. The major weaknesses in the previous techniques that applied Euclidean formulas based on

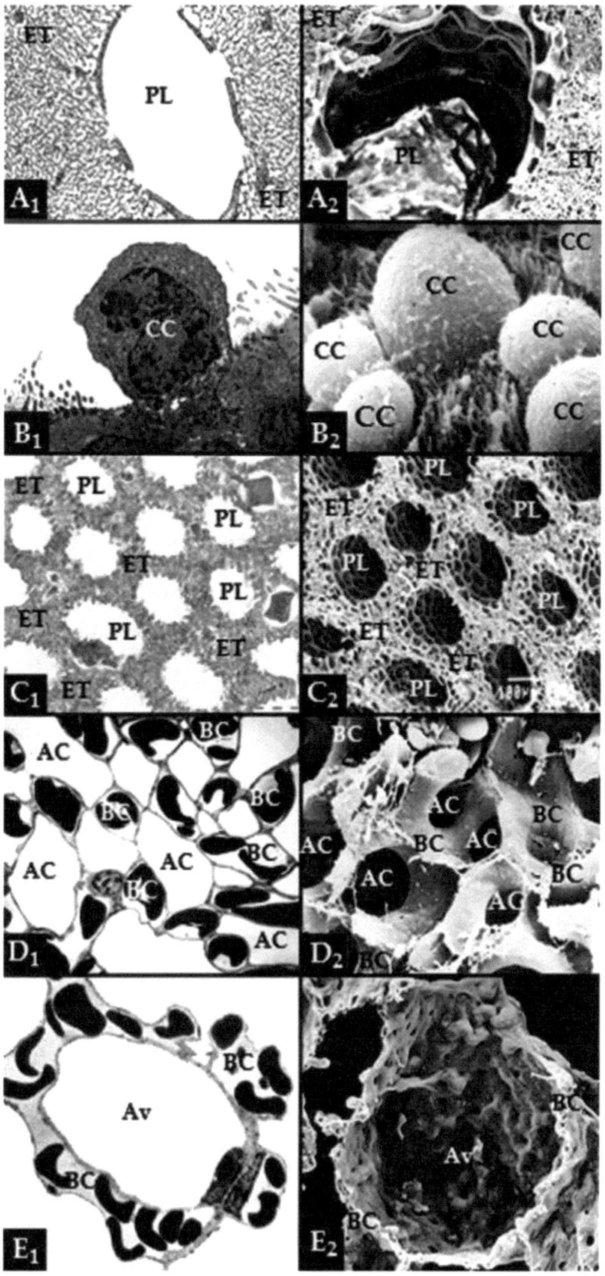

Figure 4. Sectional (2D) images (histological- or transmission electron micrographs) (A_1 to E_1) and corresponding 3D images (scanning electron micrographs) (A_2 to E_2). A_1 and A_2: Parabronchus of the lung of the ostrich, *Struthio camelus* and the domestic fowl, *Gallus gallus* variant domesticus. PL, parabronchial lumen; ET, exchange tissue; B_1 and B_2: Clara cells (CC) of the lung of the greater bush baby, *Galago senegalensis*; C_1 and C_2: Parabronchi of the lung of the domestic fowl. PL, parabronchial lumen; ET, exchange tissue; D_1 and D_2: Structural components of the exchange tissue of the lung of the domestic fowl. AC, air capillaries; BC, blood capillaries and; E_1 and E_2: Alveolus (Av) of the lung from the naked mole rat, *Heterocephalus glaber* and that of the bush baby. BC, blood capillaries. Stereology allows extrapolation of measurements made on 2D profiles (e.g. A_1 to E_1) to their 3D forms (e.g. A_2 to E_2). Absolute parameters such as volume, length, number and surface area can be determined.

classical geometric shapes were identified and improved or replaced with more robust ones. Stereology is a developing science where new innovations continue to make important improvements in the efficiency of the techniques [2, 3]. Utilizing random systematic sampling and different analytical methods, stereology provides unbiased quantitative data.

This account illustrates the versatility of stereological techniques that have contributed greatly to the better understanding of the functional design of the avian lung [6–11]. Many publications, books and programs and algorithms (softwares) are now available on the discipline. Among these are Weibel [2, 5], Stuart [12], Gundersen and Jensen [13], Gundersen and Østerby [14], Mouton [15], Howard and Reed [3] and West [16].

2. The functional design of the avian lung

2.1. Formulation of research question(s)

It is less costly for an animal to fly a given distance per unit time than it is for it to run on the ground across the same distance in the same period of time. This is notwithstanding the fact that powered (active) flight is energetically a very costly form of locomotion [17–19]. The capacity of overcoming gravity and remaining stationary in air (hovering) requires high metabolic capacity and concomitant considerably high consumption of large amounts of O_2 [20, 21]: a hovering hummingbird supports its body weight entirely by power generated by the flight muscles (**Figure 5**). Showing the highly selective nature of flight, powered flight has evolved in only a few animal taxa. It was achieved by insects ~350 million years ago (mya), by the now extinct pterodactyls ~220 mya, birds ~150 mya and bats ~50 mya, chronologically in that order. In terms of speed, endurance and high altitude travel, birds are excellent flyers. For example, a diving peregrine falcon, *Falco peregrinus*, can attain a speed of 403 kph (112 m s^{-1}) [22]; the Arctic tern, *Sterna paradisea,* flies from pole to pole, a return distance of 35,000 km [23, 24]; the American golden plover, *Pluvialis dominica,* flies 3300 km non-stop from the Aleutian Islands to the Hawaiian ones in a time of only 35 h [25]; the wandering albatross (*Diomedea exulans*) of the Southern Oceans flies continuously for days and months without landing [26, 27]; a Ruppell's griffon vulture, *Gyps rueppellii,* was sucked into the engine of a jet-craft at an altitude of 11.3 km [28]; and the bar-headed goose, *Anser indicus,* flies over some of the summits of the Himalayas, altitudes where the barometric pressure is approximately one-third of that at sea level [17, 18, 29].

Among the air-breathing vertebrates, the avian respiratory system, the so-called lung air sac system, is structurally the most complex [8, 30, 31] and functionally the most efficient [32–34].

2.2. Research question

From the above account, the research question posed was: *What are the structural adaptations, specializations and refinements that allow the avian respiratory system to acquire the large amounts of* O_2 *needed for active and sustained flight under extreme conditions such as high altitude?*

Figure 5. Active (powered) flight in a hummingbird is energetically highly costly. Insert A: In a hovering hummingbird, the body weight (white arrow) is entirely supported by power generated by the flight muscles (open arrow). Insert B. The complex avian respiratory system comprises a lung that is ventilated by air sacs. a: cervical-; b: interclavicular-, c: craniothoracic-; d: caudothoracic, and; e abdominal air sacs.

3. Stereological study of the avian lung

The health of a bird was ascertained before it was killed by sodium pentobarbitone (200 mg cm^{-3}) at a dosage of 0 $\pm$ 5 ml kg^{-1} injected into the brachial vein. The body mass was measured for normalization of data to allow intra- and inter-specific comparisons.

3.1. Fixation of the lung

A longitudinal incision was made along the neck and the trachea exteriorized. To avoid cutting any of the blood vessels associated with the neck and inadvertently introducing blood into the respiratory system, the trachea was cannulated through the larynx. The lungs were fixed by intratracheal instillation with 2.5% glutaraldehyde buffered with sodium cacodylate (pH 7.6 and osmolarity 350 mOsm) by gravity at a pressure head of 3000 Pa (1 cm H_2O = 1 mbar = 10^2 Pa). The osmolarity of the fixative was made close to the physiological one of the blood plasma [35] to avoid tissue shrinkage. Osmolarity is a critical factor in morphometric studies because hyperosmotic reagents cause tissue shrinkage, while hypo osmotic ones may cause tissue swelling. Where necessary, shrinkage factors should be determined and introduced in the final calculations. The pressure that was used to fix the lung and the air sacs have been found to be sufficient to drive the fixative into the very narrow terminal respiratory units, the air capillaries, thereby causing optimal fixation [8]. The fixation of the lung was performed by intratracheal instillation because the method preserves the erythrocytes that are important in the characterization and modelling of the lung: vascular perfusion of the lung washes away the erythrocytes. According to Crapo et al. [36], it also causes relatively greater degree of shrinkage compared to airway instillation. When the fixative stopped flowing down the trachea, the coelomic cavity was gently compressed to expel air trapped in the air sacs and that way increase the penetration of the fixative to different parts of the respiratory system. The trachea was then ligated and the fixative left *in situ* for ~4 h. Thereafter, the lungs were carefully removed from their costovertebral attachments and immersed in fixative.

3.2. Determination of the lung volume

The extrapulmonary primary bronchus was trimmed at the hilum of the lung and the adhering fat and connective tissue removed before the volume of the lung was determined. This was done by weight displacement method [37] (**Figure 6**) which is based on Archimedes' principle that states that a floating body displaces its own weight. Since the specific gravity of the water is unity, the increase in the weight caused by the volume of the fluid displaced by freely

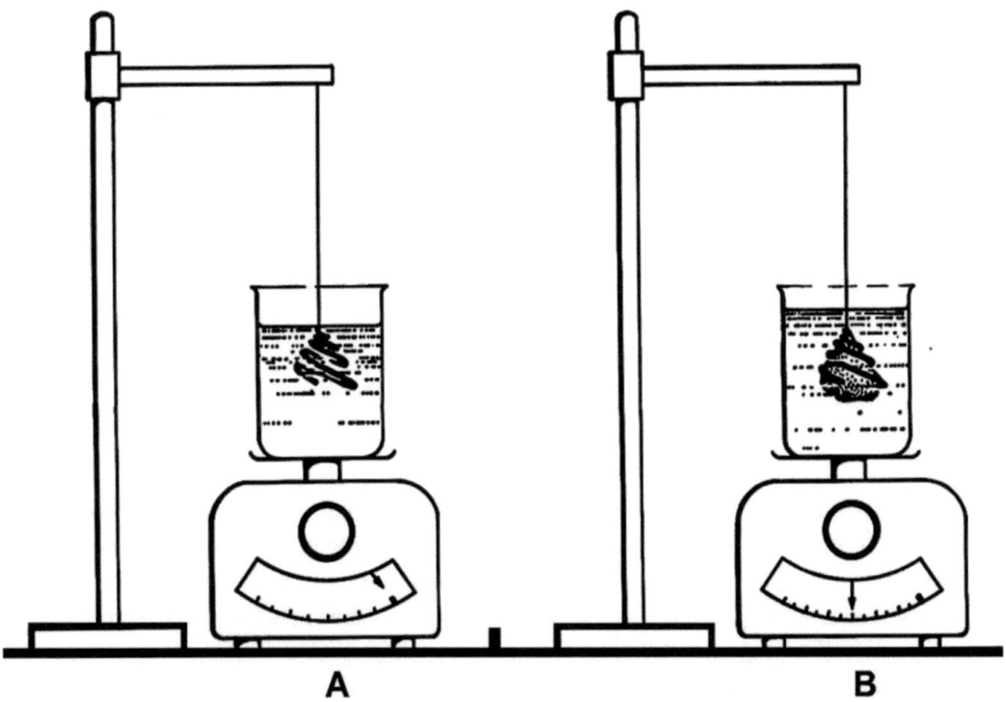

Figure 6. Determination of lung volume by Scherle's method [37] which is based on the Archimede's principle. Because the specific gravity of the lung is less than one and it therefore floats on water, a piece of wire (A) of known volume was used to submerge the lung (B). After the balance was tarred to zero (A) and the lung freely suspended in water (B), the weight increase was equivalent to the volume of the lung because the specific gravity of water is unity.

suspended lung equals its volume. As the specific gravity of the lung is less than one and hence the lung floats to the surface of the water, a metal wire of known volume was used to keep the lung submerged under water. At least three close measurements were made, and the average volume of the lung calculated.

Scherle's [37] method is very accurate in determining volumes of small objects. There being no morphological and morphometric differences between the left and the right avian lungs [8], the left lung was used for light microscopic analyses and the right one for electron microscopy.

3.3. Light microscopic analysis

3.3.1. Tissue processing and sampling

Very small lungs were processed by the standard laboratory techniques, embedded in paraffin wax and serial transverse sections cut craniocaudally at a thickness of 10 μm. The sections were then stained with haematoxylin and eosin. Eight equidistantly spaced, i.e. stratified, sections were taken for analysis. The lungs were cut into slices along the costal sulci and the slices were then cut into halves just dorsal to the primary bronchus (**Figure 7**). Facing cranially, the half slices were processed and embedded in paraffin wax. The first technically adequate section from the cranial face of each half slice was stained with haematoxylin and eosin for analysis.

3.3.2. Determination of volume densities

The volume densities, i.e. the fractional or proportional volumes, of the exchange tissue, the lumina of the parabronchi and secondary bronchi, the blood vessels larger than blood capillaries and the primary bronchus were determined by point-counting at a magnification of 100× using a 100-point Zeiss integrating graticule (**Figure 8**). For example, for the exchange tissue, the volume density ($V_{V(ET)}$) was calculated as follows:

$$V_{V(ET)} = P_{(ET)}.P_T^{-1} \tag{1}$$

where $P_{(ET)}$ is the number of points falling onto the exchange tissue and P_T the total number of points in the test system or the reference space.

The absolute volumes of the structural components were calculated from the volume of the lung (V_L). For example, the absolute volume of the exchange tissue (V_{ET}) was calculated as follows:

$$V_{ET} = V_{V(ET)}.V_L \tag{2}$$

The volume densities of the components of the lung that comprise insignificant volume of the lung, e.g. the lymphatics and the inter-parabronchial septa, were not determined. The parabronchi and the secondary bronchi were combined because most of the secondary bronchi have a gas exchange tissue mantle surrounding them, and the small secondary bronchi cannot be well differentiated from the parabronchi.

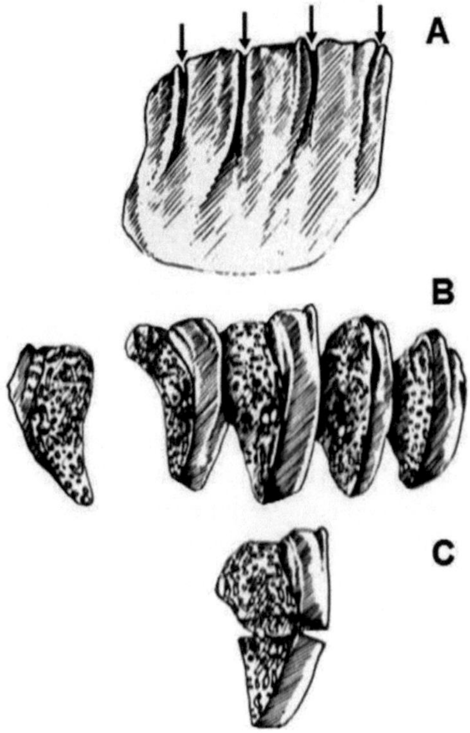

Figure 7. Stratified sampling of the avian lung. The lung (A) was cut into slices along the costal sulci (arrows) (A) and the slices (B) cut into halves just above the primary bronchus (C).

3.3.3. Sample size sufficiency

The adequacy of the number of sections analysed was determined in a pilot study by plotting cumulative average graphs on analysis made on stratified sections, and the sufficiency of the number of points counted for a particular structural component from nomograms given in Weibel [2]. Since the sections were analysed entirely, i.e. field by field, the number of points counted for the three main structural components, i.e. the exchange tissue, the lumina of the parabronchi and the secondary bronchi and the blood vessels larger than blood capillaries, surpassed those needed to give a standard error of the mean of 5% or less.

3.4. Electron microscopic analysis

3.4.1. Tissue processing and sampling

The right lung was cut into slices along the costovertebral sulci and the slices were then cut into halves dorsal to the primary bronchus. The half slices were diced, and the pieces (~1 mm^3 in size) were processed for electron microscopy by the standard laboratory techniques (**Figure 9**). From each half slice, 4–10 blocks were prepared. One block was picked at random and trimmed to remove the rest of the structural components, leaving only the exchange

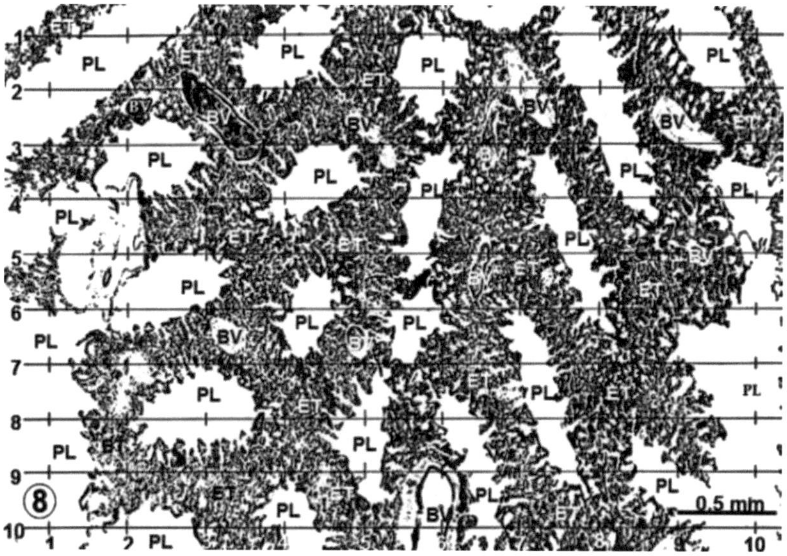

Figure 8. A histological section of the avian lung stained with haematoxylin and eosin. A Zeiss integrating graticule with 100 points was superimposed onto the section. The structural components were analysed by point-counting. PL, parabronchial lumen; ET, exchange tissue; BV, blood vessel larger than blood capillaries.

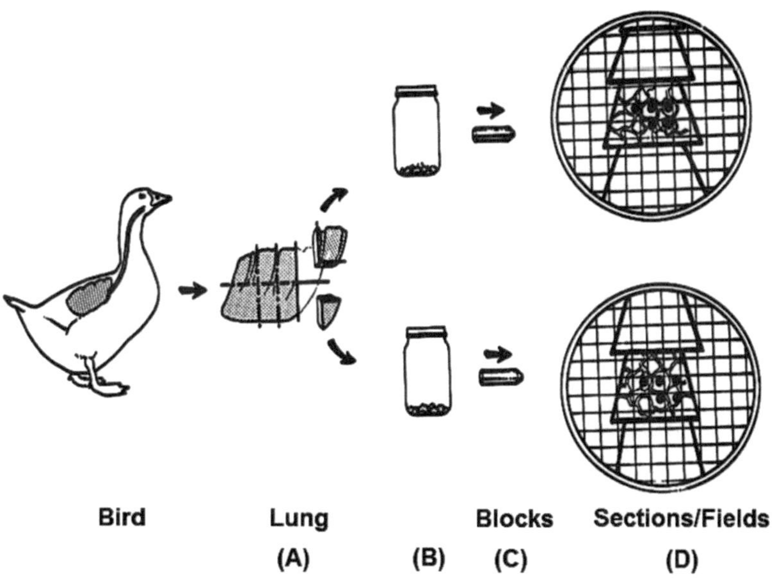

Figure 9. Stratified sampling of the avian lung for transmission electron microscopic analysis. The right lungs were cut transversely along the costal sulci and the slices then cut into halves (A). The halve slices were then diced into small pieces (~1 mm^3) which were embedded in epon (B). From a number of blocks prepared from a half slice, a block (C) was picked at random and ultrathin sections cut and mounted onto 200-wire mesh grids (D). Electron micrographs taken from predetermined areas, i.e., the top right-hand corner of the grid squares (D), to avoid bias.

tissue. Ultrathin sections were cut and mounted on 200-square wire mesh copper grids. Five micrographs were taken from a predetermined corner of the grid squares (the top right corner) to avoid bias at a primary magnification of 3000×. The images were enlarged by a factor of 2.5 and a quadratic lattice grid superimposed on the image (**Figure 10**). On average, for each bird, a total of 40 electron micrographs were analysed at a final magnification of 7500×. In a pilot study, the magnification used for the analysis provided a large field of investigation while providing adequate resolution and permitting the counts and the measurements to be made accurately.

3.4.2. Determination of volume densities

The volume densities of the components of the exchange tissue, namely, the air capillaries, the blood capillaries, the supportive tissue, i.e. the tissue of the blood-gas barrier and the parts of the exchange tissue that are not involved in gas exchange, and the erythrocytes were determined by point counting (**Figure 10**). The intersections of the vertical and the horizontal lines constituted the points used for the determination of the volume densities. For example, the volume density of the air capillaries ($V_{V(ac)}$) was determined as follows:

$$V_{V(AC)} = P_{(AC)}.P_{T}^{-1} \tag{3}$$

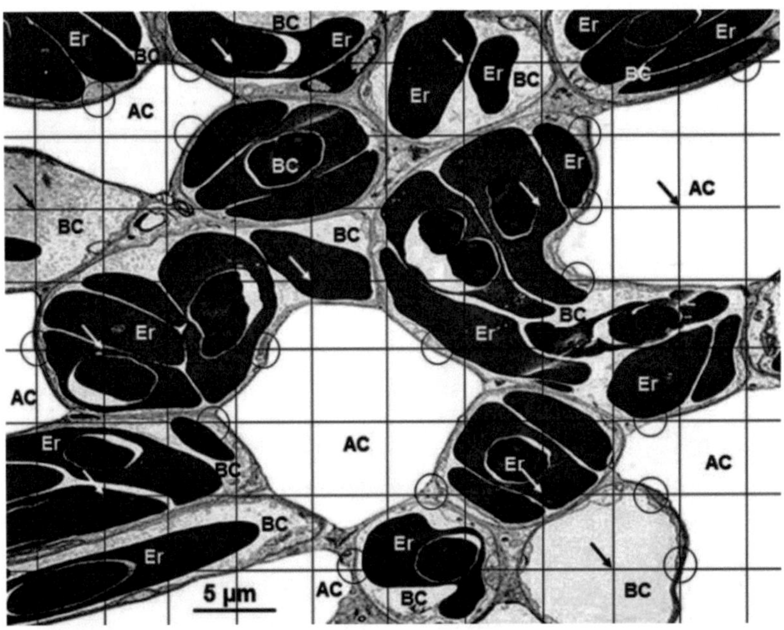

Figure 10. An electron micrograph of the exchange tissue of the avian lung onto which a quadratic lattice grid is superimposed. Points that were formed by the intersections between the vertical and the horizontal lines (arrows) were used to determine the volume densities of the air capillaries (AC), the blood capillaries (BC), the erythrocytes (Er) and the supporting tissue, i.e. the tissue of the blood gas (tissue) barrier and the tissue not involved in gas exchange. The intersections of the horizontal lines with the tissue barriers (circles) were used to determine the surface areas of the air capillaries, the capillary endothelium and the erythrocyte cell membrane.

where $P_{(AC)}$ is the number of points falling onto the air capillaries and P_T the total number of points in the test system.

The absolute volume of the air capillaries (V_{AC}) was calculated from its volume density ($V_{V(AC)}$) and the volume of the exchange tissue (V_{ET}) as follows:

$$V_{(AC)} = V_{V(AC)}.V_{(ET)}^{-1} \tag{4}$$

3.4.3. Determination of surface densities and surface areas

The surface densities of the blood-gas (tissue) barrier, the blood capillary endothelium and the cell membrane of the erythrocytes were determined by intersection counting, i.e. by counting the crossings of the test system, i.e. the horizontal lines of a quadratic lattice grid, with particular tissue barrier (**Figure 10**). In a pilot study, it had been determined that the number of vertical and horizontal intersections with the barriers were not statistically different. This showed that the

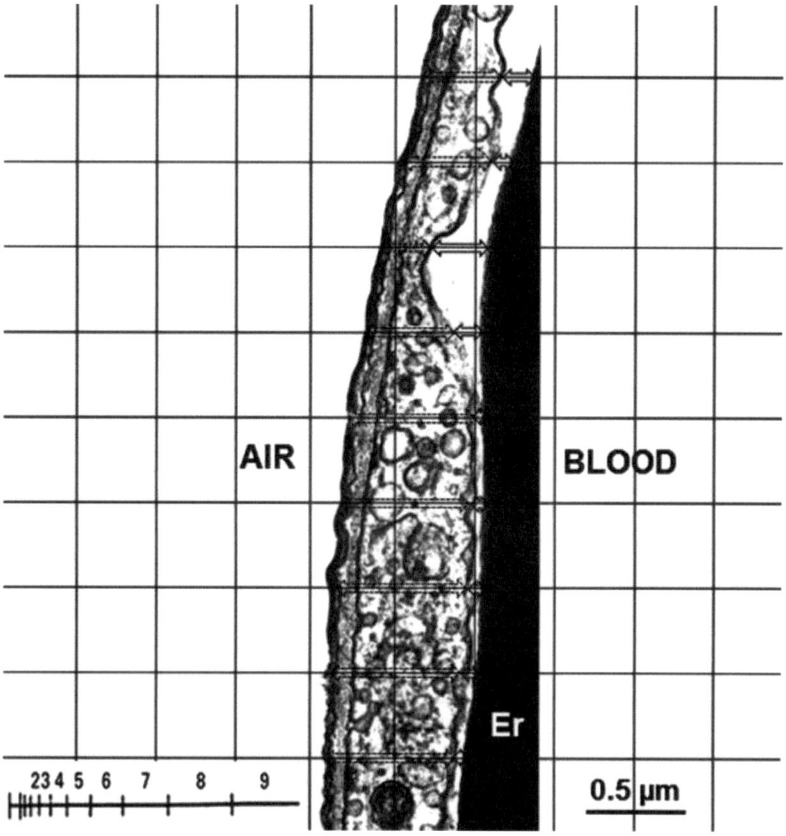

Figure 11. The harmonic thicknesses of the tissue barrier and the plasma layer were determined by intercept length measurement along the horizontal lines of the test grid. The dashed double sided arrows show the intercepts of the blood-gas (tissue) barrier while the continuous double sided ones show those of the plasma layer. A logarithmic scale was used to measure the harmonic mean thickness of the blood-gas (tissue) barrier and the plasma layer. It is shown on the bottom left corner of the figure.

exchange tissue of the avian lung is anisotropic, i.e. homogeneous. For example, the surface density of the blood-gas (tissue) barrier (BGB) ($S_{V(BBG)}$) was calculated as follows:

$$S_{V(BBG)} = 2I_{Lt} \quad (5)$$

where I is the number of intersections and Lt the total length of the test system in real units, i.e. after correction for the magnification.

The surface area of the blood-gas (tissue) barrier ($S_{A(BGB)}$)was calculated as the product of its surface density (S_V) and the volume of the exchange tissue (V_{ET}) as follows:

$$S_{A(BGB)} = S_{V(BGB)} . V_{(ET)} \quad (6)$$

3.4.4. Determination of harmonic mean thickness

The harmonic mean thicknesses of the blood-gas (tissue) barrier (τht) and the plasma layer (τhp) were determined from the sum of the reciprocals of the respective intercept lengths (l_h) measured using a logarithmic scale (**Figure 11**). Harmonic mean thickness and NOT arithmetic thickness that weighs the smaller intercepts (thicknesses) compared to the larger ones is the more appropriate estimator of the diffusing capacity, i.e. the conductance, of a barrier to O_2. τht and τhp were calculated as follows:

$$\tau ht = \tfrac{2}{3} \cdot l_h \quad (7)$$

where l_h is the mean intercept length measured on a logarithmic scale. The mean intercept length was divided by the final magnification to express the thickness in real units.

4. Pulmonary modelling

4.1. General considerations

Following a hierarchy that can be examined on a scale from microscopic to gross, living things are highly organized and structured entities. In larger organisms, cells combine to comprise tissues, which are groups of similar cells performing similar or related functions. Organs are collections of tissues grouped together executing a common function to meaningfully explicate how and why animals work the way they do, biologists must adopt appropriate conceptual models of engineers. On their own, quantitative data do not quite explain the function of an organism or that of its constituent parts. Interestingly, the sum total of the functions performed by the different parts of an organism, e.g. organelles, cells, tissues, organs and organ systems, surpass the function expressed by the whole organism [38, 39]. Hammond et al. [40] stated that 'natural selection operates on organismal-level traits that are usually manifestations of the integrated functioning of a suite of organs and organ systems'. In the complex dynamic biological entities, measurement of physiological changes and processes by testing different structures and measuring separate functions therefore leads to wrong deductions.

Like other organs and organ systems, gas exchangers possess a complex cascading assemblage of structural components that span from cellular- to organ-system level: the structural components are organized as discrete but functionally integrated units. For respiratory organs, the physiological process (gas exchange) that manifests at organismal level is a product of infinitely many small events that are generated by various structural entities at the different levels of organization. Gans [41, 42] pointed out that animals very rarely exhibit one-function-one-structure designs: 'each activity tends to involve multiple aspects of the phenotype and each aspect of the phenotype may be involved in multiple activities'.

Comparative respiratory biologists focus on the structure of gas exchangers and determine how they correlate with properties like function, phylogeny, environment, body size and lifestyle. An insightful study of a gas exchanger or for that matter any other organ should involve application of comprehensive physical models that mathematically integrate the structural and functional aspects of the whole organism/animal.

4.2. Biological models

Scheid [43] remarked that 'models are not only helpful but often indispensable in quantitative biology' and that a model is 'an image of part of the physical or conceptual world apt to explain or predict observations'. Gutman and Bonik [44] termed a model 'an abstraction of a real situation that describes only the essential aspects of the situation'. A mathematical model separates a complex biological structure into its functional parts, sets apart those that are most important in answering particular research questions and then integrates them. A biological model is a mathematical simplification of a complex structure that satisfactorily characterizes the system it describes. It should be simple to apply, easy to understand and theoretically and practically testable.

A simple mechanistic model of a gas exchanger consists of a structure in which the external and the internal respiratory media are separated by a physical (tissue) barrier across which partial pressure gradient of oxygen (PO_2) exists (**Figure 12**). Powell and Scheid [45] pointed out that 'in an attempt at deriving a functional model for gas exchange from morphologic evidence, the physiologist has to identify the simplest functional subunit in the gas exchange organ'. Although biological models are mathematical abstractions of complex systems, since the functions of organs and organisms are regulated by many variables, models should be highly instructive in comparative studies where similar functions are performed by various structures in different ways. To simplify biological models, many structural details have to be omitted and certain assumptions made. The exclusions and the conjectures determine the predictive power of a model. Tweaking a model provides information on the relationship between the physical and functional parameters that drive a biological system. Mathematically adjusting one or more of the parameters and the prevailing conditions under which a system works while holding others constant allows for the identification of constraining, potentiating and redundant factors. The underlying multifaceted control mechanisms that drive the performance of biological components, systems and whole organisms can best be identified by careful modelling.

Figure 12. The main parts of a gas exchanger. Oxygen (O_2) diffuses across a tissue barrier under a partial pressure gradient (large arrow). The conductance of the barrier to O_2 (Dto_2) correlates directly with the surface area (S) and inversely with the thickness of the barrier (t): equation.

4.3. Morphological basis of pulmonary modelling

In accordance with the Fick's law, the conductance or the volume of a gas (e.g. oxygen) that is transferred by diffusion across a tissue barrier per unit time (Do_2) is directly proportional to the surface area (S), the Krogh's permeation coefficient across the tissue barrier (Kto_2) and the partial pressure gradient of O_2 (ΔPo_2). Do_2 correlates inversely with the thickness of the blood-gas (tissue) barrier (t), i.e. the distance O_2 molecules diffuse (**Figure 12**). Fick's law is expressed as follows:

$$Do_2 = Kto_2 \cdot S \cdot \Delta Po_2 \cdot T^{-1} \tag{8}$$

It is the physiologist's equivalent of Ohm's law of electricity which is expressed as follows:

$$I = U \cdot R^{-1} \tag{9}$$

where I is the electric current, U the potential difference (i.e. the voltage) and R the resistance.

The morphometric diffusing capacities of the components of the lung can be estimated from the respiratory surfaces area, the thickness of the air-haemoglobin pathway, the volume of blood in the blood capillaries, the Krogh's O_2 permeation coefficients and the O_2 uptake coefficient of the whole blood [46, 47].

To various extents, the respiratory organs (gas exchangers) have been morphometrically analysed and functionally modelled. These include the fish gills [48] and the lungs of the lungfish (Dipnoi) [49], reptiles [50–53], birds [8, 54–57] and mammals [58–61]. From the concepts and findings of Roughton [62], Roughton and Forster [63], Staub et al. [64] and Sackner et al. [65], a morphometric model was developed by Weibel [66] and later revised in Weibel et al. [67] (**Figure 13**). The relationship between the total pulmonary diffusing capacity of the lung for O_2 (DLo_2) and that of its other parts, namely, the membrane-diffusing capacity (Dmo_2) and the erythrocyte (Deo_2) relate as follows:

$$DLo_2{}^{-1} = Dmo_2{}^{-1} + Deo_2{}^{-1} \tag{10}$$

Deo_2 is the product of Θo_2,the binding rate of O_2 to haemoglobin and the pulmonary capillary blood volume (Vc; Eq. (12)).

Although in all gas exchangers O_2 diffuses across the so-called air-haemoglobin pathway that essentially comprises the blood-gas (tissue) barrier, the plasma layer and to a certain extent, the cytoplasm of the erythrocyte before the molecule is biochemically bound to the haemoglobin, certain modifications of the basic model (**Figures 12** and **13**) have been necessary to satisfy the variations in the morphologies of the different respiratory organs and structures. For example, in birds where the erythrocytes are nucleated, the volume of the pulmonary capillary blood has to be adjusted by 'subtracting' the volume occupied by the nuclei in the erythrocytes [7, 8, 56]. The diffusing capacities of the blood-gas (tissue) barrier (Dto_2) and the plasma layer (Dpo_2) are estimated from their respective surface areas (S), their harmonic mean thicknesses (τh) and their Krogh's permeation constants (K) for O_2(Ko_2). For example, for the blood-gas (tissue) barrier, the diffusing capacity of the barrier (Dto_2) is calculated as follows:

$$Dto_2 = Kto_2 \cdot St \cdot \tau ht^{-1} \tag{11}$$

where Kto_2 is the Krogh's O_2 permeation constant through the blood-gas (tissue) barrier, St is the surface area of the blood-gas (tissue) barrier and τht is the harmonic mean thickness of the blood-gas (tissue) barrier.

The quantity of blood in the blood capillaries of a respiratory organ determines the amount of O_2 bound by the haemoglobin [63, 68]. The diffusing capacity of the erythrocytes (Deo_2) is calculated as follows:

$$Deo_2 = Vc \cdot \Theta o_2 \tag{12}$$

where Vc is the volume of the pulmonary capillary blood and Θo_2 the binding rate of O_2 to haemoglobin. The O_2 permeation constant (K) is a product of the solubility (a) and diffusion (D) constants. Since temperature affects the two factors in opposite directions, i.e. solubility decreases while diffusion increases, Kto_2 is not significantly affected by change in temperature.

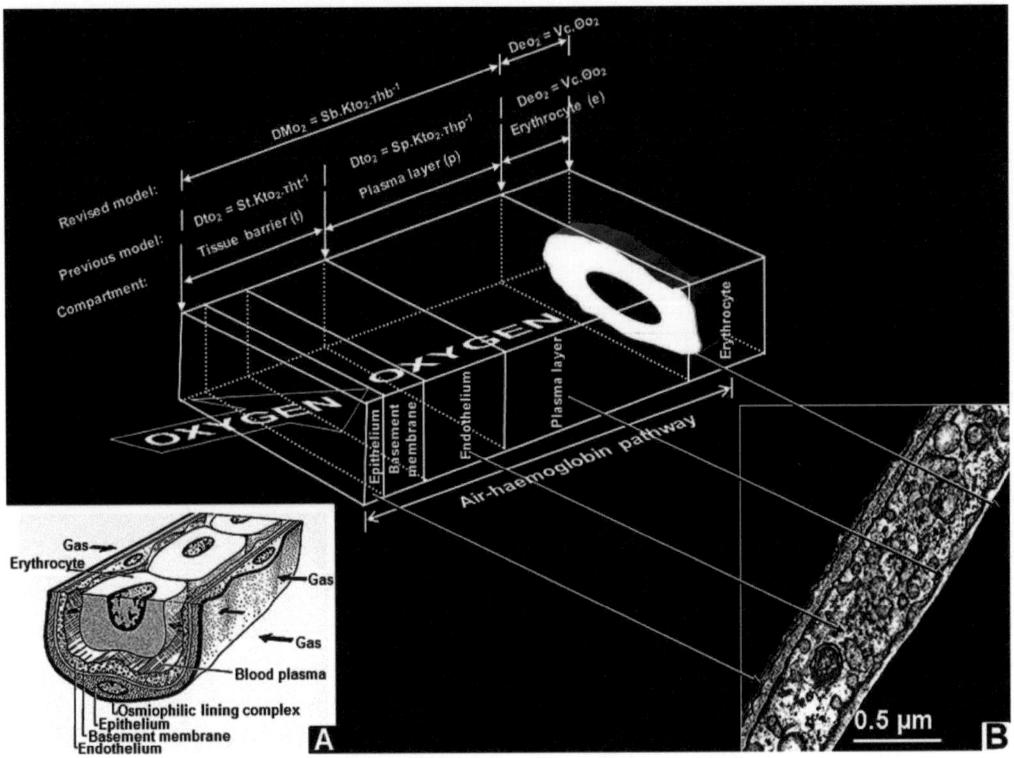

Figure 13. A stereogram showing that in a gas exchanger, oxygen (O_2) diffuses under a partial pressure gradient (large arrow marked oxygen). The barriers through which O_2 diffuses, the so-called the air-haemoglobin pathway, comprises the blood-gas (tissue) barrier, the plasma layer and the cytoplasm of the erythrocyte (RBC). The blood-gas (tissue) barrier consists of an epithelial cell, a basement membrane and an endothelial cell. According to the previous model of Weibel [66], the conductance, i.e., the diffusing capacity of the blood-gas (tissue) barrier for O_2 (Dto_2) is calculated from the surface area of the barrier (St), the O_2 permeation constant through the tissue barrier (Kto_2) and the harmonic mean thickness of the barrier (τht); the conductance of the plasma layer for O_2 (Dpo_2) is calculated from the surface area of the plasma layer (Sp), the O_2 permeation constant through the plasma layer (Kpo_2) and the harmonic mean thickness of the plasma layer (τhp); the conductance of the erythrocyte for O_2 (Deo_2) is calculated from the volume of the pulmonary capillary blood (Vc) and the O_2 uptake coefficient (θo_2). The diffusing capacities correlate directly with the surface areas (S) and the O_2 permeation coefficients (K) and inversely with the thicknesses of the barriers. In the revised model of Weibel et al. [67], the thicknesses of the blood-gas (tissue) barrier and the plasma layers are combined to form the harmonic mean thickness of the total barrier (τhb) and used to calculate the membrane diffusing capacity (Dmo_2). The total (overall) pulmonary morphometric diffusing capacity is calculated from the reciprocals of Dto_2, Dpo_2 [or DMo_2] according to the revised model of Weibel [67] and the Deo_2. Inserts: Insert A: Diagram showing a cross section of a blood capillary which is opened to show erythrocytes and gas (O_2) diffusing through the blood-gas (tissue) barrier and the plasma layer before being bound to the haemoglobin. Insert B: Transmission electron micrograph showing the blood-gas (tissue) barrier and the plasma layer (the air-haemoglobin pathway) of the lung of the domestic fowl, *Gallus gallus* variant *domesticus*.

The barriers that form the air-haemoglobin pathway are arranged in series, i.e. an O_2molecule has to pass through these barriers in succession before it binds to the haemoglobin. Like for electricity when the resistances are arranged in series, in the lung, the total resistance that the molecule encounters can be mathematically expressed as follows:

$$R_L = R_t + R_p + R_e \tag{13}$$

where R_L is the total resistance the lung confers to O_2 molecules and R_t, R_p and R_e are respectively the resistances offered by the blood-gas (tissue) barrier, the plasma layer and the erythrocyte.

According to the 'older' model of Weibel [66], the membrane-diffusing capacity ($D\text{mo}_2$) is the combined diffusing capacity (conductance) of the blood-gas (tissue) barrier ($D\text{to}_2$) and the plasma layer ($D\text{po}_2$) and is calculated as follows:

$$D\text{mo}_2 = D\text{to}_2 + D\text{po}_2 \tag{14}$$

In the revised model of Weibel et al. [67], the thickness of the plasma layer is combined with that of the blood-gas (tissue) barrier so that $D\text{mo}_2$ is calculated as follows:

$$D\text{mo}_2 = S_{(\text{tb})} \cdot \tau\text{hb} \tag{15}$$

where $S_{(\text{tb})}$ is the total respiratory surface area and τhb is the harmonic mean thickness of the total barrier, i.e. the distance from the respiratory surface to the cell membrane of the erythrocyte.

The total morphometric pulmonary diffusing capacity ($D\text{Lo}_2$) is determined from the diffusing capacities of the blood-gas (tissue) barrier ($D\text{to}_2$), the plasma layer ($D\text{po}_2$) and that of the erythrocytes ($D\text{eo}_2$) as follows:

$$D\text{Lo}_2 = D\text{to}_2 + D\text{po}_2 + D\text{eo}_2 \tag{16}$$

$D\text{Lo}_2$ is an integrative parameter that expresses the structural capacity of the lung to transfer (conduct) oxygen to the body.

4.4. Significance of mathematical modelling in biology

Inexperience and/or unawareness on the technique and lack of physical constants of O_2 permeability through tissues have particularly hindered investigators from mathematically modelling respiratory organs. Comparisons of the functional designs of gas exchangers have largely been based on relating single structural parameters such as lung volumes, respiratory surface areas and thicknesses of the blood-gas barriers. In some cases [50–52], the so-called 'anatomical diffusion factor' (ADF) which is the ratio of respiratory area to the thickness of the blood-gas (tissue) barrier has been used to assess functional efficiencies. Using individual morphometric parameters can lead to wrong conclusions. For example, among birds on which pulmonary morphometric data exist, the Humboldt penguin, *Spheniscus humboldti,* was reported to have a particularly thick blood-gas (tissue) barrier of a harmonic mean thickness of 0.530 μm [57]. If the harmonic mean thickness of the blood-gas (tissue) tissue barrier was used to compare the efficiency of the penguin's lung with those of other birds, it would have been concluded that the gas exchange efficiency of the penguin lung is very poor. However, because like in other diving animals the volume of blood in the lung is very large [69], the total morphometric pulmonary diffusing capacity of the lung of the Humboldt penguin corresponds with those of other species of birds of equivalent body mass [57]. Both in normal and pathological states, changes in the diffusing capacity of the lung can occur anywhere along the air-haemoglobin pathway. In conditions such as pulmonary edema, the thickness of the blood-gas (tissue) barrier increases; in atelectasis (collapse of the lung), the respiratory surface area decreases; in emphysema, respiratory surface area decreases because of damage of the interalveolar septa; and in some cases of anaemia, the volume of the erythrocytes and therefore the quantity of haemoglobin decreases.

Application of integrative mathematical models on data acquired from biological structures provides robust answers to research questions.

5. Finding

Quantitative analyses of lungs of different species of birds have shown that pulmonary structural refinements correspond with factors such as body mass, lifestyle and habitat occupied [7, 8, 56] (**Figure 14**).

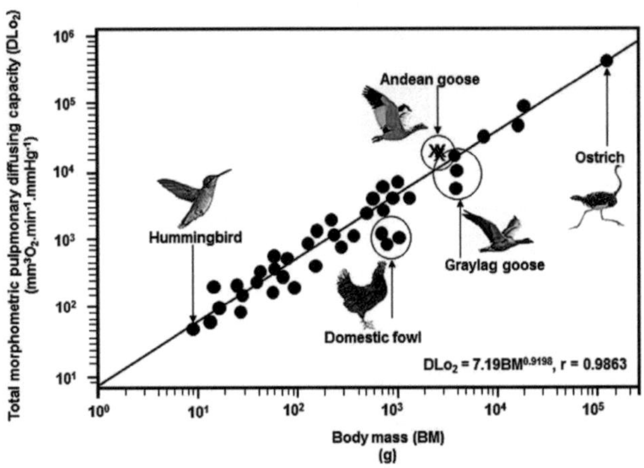

Figure 14. Regression line showing the correlation between the total morphometric pulmonary diffusing capacity (DLo_2) of birds against body mass (BM). Some of the species of birds that have been studied are shown. The data on which the regression line was plotted are given in Maina [7, 8] and Maina et al. [56].

Acknowledgements

The writing of this chapter was supported by the National Research Foundation of South Africa (NRF). The views and opinions expressed here are, however, those of the author and are not necessarily of the NRF.

Author details

John N. Maina

Address all correspondence to: jmaina@uj.ac.za

Department of Zoology, University of Johannesburg, Johannesburg, South Africa

References

[1] Guex J. Retrograde evolution during major extinction crises. In: Verkhratsky A, editor. Springer Briefs in Evolutionary Biology. Berlin: Springer; 2015. pp. 1–74

[2] Weibel ER. Stereological Methods, Vol. 1: Practical Methods for Biological Morphometry. London: Academic Press; 1979

[3] Howard CV, Reed MG. Unbiased Stereology. 2nd ed. Liverpool: QTP Publications; 2010

[4] Elias H. Address of the president. In: Haug H, editor. Proceedings of the 1st International Congress for Stereology. Wien: Congressprint; 1963. p. 2

[5] Weibel ER. Stereological methods, Vol. 2: Practical Methods for Biological Morphometry. London: Academic Press; 1980

[6] Maina JN. The Gas Exchangers: Structure, Function and Evolution of the Respiratory Processes. Heidelberg: Springer; 1998

[7] Maina JN. The morphometry of the avian lung. In: King AS, McLelland J editors. Form and Function in Birds. Vol. 4. London: Academic Press; 1989; pp. 307–368

[8] Maina JN. The Lung-Air Sac System of Birds: Development, Structure and Function. Heidelberg: Springer; 2005

[9] Maina JN. Development, structure and function of a novel respiratory organ, the lung-air sac system of birds: To go where no other vertebrate has gone. Biological Reviews. 2006;**81**:545–579

[10] Maina JN. The design of the avian respiratory system: Development, morphology and function. Journal of Ornithology. 2015;**156**:41–63

[11] Maina JN, West JB. Thin but strong! The dilemma inherent in the structural design of the blood-water/gas barrier: Comparative functional and evolutionary perspectives. Physiological Reviews. 2005;**85**:811–844

[12] Stuart A. Basic Ideas of Sampling. London: Griffin and Co.; 1984

[13] Gundersen HJG, Jensen EB. The efficiency of systematic sampling in stereology and its prediction. Journal of Microscopy. 1987;**147**:229–262

[14] Gundersen HJG, Østerby R. Optimizing sampling efficiency of stereological studies in biology: Or do more less well. Journal of Microscopy. 1981;**121**:65–73

[15] Mouton PR. Principles of Unbiased Stereology: An Introduction for Bioscientists. Baltimore (MD): John Hopkins University Press; 2002

[16] West MJ. Basic Stereology for Biologists and Neuroscientists. 1st ed. New York: Cold Spring Harbor Laboratory Press; 2012

[17] Bishop CM, Butler PJ. Flight. In: Scanes CG, editor. Sturkie's Avian Physiology. 6th ed. New York: Academic Press; 2015. pp. 919–974

[18] Butler PJ. High fliers: The physiology of bar-headed geese. Comparative Biochemistry and Physiology A. 2010;**156**:325–329

[19] Butler PJ. The physiological basis of bird flight. Philosophical Transactions of the Royal Society B. 2016;**371**:20150384

[20] Bartholomew GA, Lighton JRB. Oxygen consumption during hover-feeding in free-ranging Anna hummingbirds. Journal of Experimental Biology. 1986;**123**:191–199

[21] Wells DJ. Muscle performance in hovering hummingbirds. Journal of Experimental Biology. 1993;**78**:39–57

[22] Tucker VA. Gliding flight: Speed and acceleration of ideal falcons during diving and pull out. Journal of Experimental Biology. 1998;**201**:403–414

[23] Salomonsen F. Migratory movements of the Arctic tern (*Sterna paradisea pontoppidan*) in the Southern Ocean. Det Kongelige Danske Videnskabernes Selskab Biologie Medicine. 1967;**24**:1–37

[24] Egevang C, Stenhouse IJ, Phillips RA, Petersen A, Fox JW, Silk JR. Tracking of Arctic terns, *Sterna paradisea* reveals longest animal migration. Proceedings of the National Academy of Science United State of America. 2010;**107**(5):2078–2081

[25] Johnston DW, McFarlane RW. Migration and bioenergetics of flight in the Pacific golden plover. Condor. 1967;**69**:156–168

[26] Lockley RM. The most aerial bird in the world. Animals. 1970;**13**:4–7

[27] Weimerskirch H, Delord K, Guitteaud A, Phillips RA, Pinet P. Extreme variation in migration strategies between and within wandering albatross populations during their sabbatical year, and their fitness consequences. Scientific Reports. 2015;**5**:8853

[28] Laybourne RC. Collision between a vulture and an aircraft at an altitude of 37,000 ft. Wilson Bulletin. 1974;**86**:461–462

[29] Scott GR, Hawkes LA, Frappell PB, Butler PJ, Bishop CM, Milsom WK. How bar-headed geese fly over the Himalayas. Physiology. 2015;**30**:107–115

[30] King AS. Structural and functional aspects of the avian lung and its air sacs. International Reviews of General Experimental Zoology. 1966;**2**:171–267

[31] Duncker HR. The lung-air sac system of birds. A contribution to the functional anatomy of the respiratory apparatus. Ergebnesse Anatomie Entwicklung. 1971;**45**:1–171

[32] Scheid P. Mechanisms of gas exchange in bird lungs. Reviews of Physiology, Biochemistry and Pharmacology. 1979;**86**:137–186

[33] Fedde MR. The structure and gas flow pattern in the avian lung. Poultry Science. 1980;**59**:2642–2653

[34] Powell FL. Respiration. In: Scanes C, editor. Sturkie's Avian Physiology. 6th ed. San Diego: Elsevier; 2015. pp. 301–336

[35] Sykes AH. Formation and composition of urine. In: Bell DJ, Freeman BM, editors. Physiology and Biochemistry of the Domestic Fowl. London: Academic Press; 1971. pp. 233–278

[36] Crapo JD, Crapo RO, Jensen RL, Mercer RR, Weibel ER. Evaluation of lung diffusing capacity by physiological and morphometric techniques. Journal of Applied Physiology. 1988;**64**:2083–2091

[37] Scherle WF. A simple method for volumetry of organs in quantitative stereology. Mikroskopie. 1970;**26**:57–60

[38] Thompson D'AW. On Growth and Form. 2nd ed. Cambridge: Cambridge University Press; 1959

[39] Hoagland M, Dodson B. The Way Life Works. London: Ebury Press; 1995

[40] Hammond KA, Chappell MA, Cardullo RA, Lin RS, Johnsen TS. The mechanistic basis of aerobic performance variation in red jungle fowl. Journal of Experimental Biology. 2000;**203**:2053–2064

[41] Gans C. Vertebrate morphology: Tale of a phoenix. American Zoologist. 1985;**25**:689–694

[42] Gans C. Adaptation and the form-function relation. American Zoologist. 1988;**28**:681–697

[43] Scheid P. The use of models in physiological studies. In: Feder ME, Bennett AF, Burggrenn WW, Huey RB, editors. New Direction in Ecological Physiology. Cambridge: Cambridge University Press; 1987. pp. 275–288

[44] Gutman WF, Bonik K. Kritische Evolutionstheorie. Gerstenberg: Hildesheim; 1981

[45] Powell FL, Scheid P. Physiology of gas exchange in the avian respiratory system. In: King AS, McLelland J, editors. Form and Function of the Avian Lung. Vol. 4. London: Academic Press; 1989. pp. 393–437

[46] Vandergriff KD, Olson JS. Morphological and physiological factors affecting oxygen uptake and release by red blood cells. Journal of Biological Chemistry. 1984;**259**:12619–2627

[47] Yamaguchi K, Nguyen-Phu D, Scheid P, Piiper J. Kinetics of oxygen uptake and release by human erythrocytes studied by a stopped-flow technique. Journal of Applied Physiology. 1985;**58**:215–1224

[48] Hughes GM. Morphometrics of the fish gills. Respiration Physiology. 1972;**14**:1–25

[49] Hughes GM, Weibel ER. Morphometry of fish lungs. In: Hughes GM, editor. Respiration of Amphibious Vertebrates. London: Academic Press; 1976. pp. 213–232

[50] Perry SF. Quantitative anatomy of the lungs of the red-eared turtle, *Pseudemys scripta elegans*. Respiration Physiology. 1978;**35**:245–262

[51] Perry SF. Morphometric analysis of pulmonary structure: Methods for evaluation and comparison of unicameral lungs. Microskopie. 1981;**38**:278–293

[52] Perry SF. Reptilian lungs: Functional anatomy and evolution. Advances in Anatomy Embryology and Cell Biology. 1983;**79**:1–81

[53] Perry SF. Recent advances and trends in the comparative morphometry of vertebrate gas exchange organs. In: Boutilier RG, editor. Advances in Comparative and Environmental Physiology. Heidelberg: Springer-Verlag; 1990. pp. 45–71

[54] Maina JN. Functional morphology of the avian respiratory system, the lung-air sac system: Efficiency built on complexity. Ostrich. 2008;**79**:117–132

[55] Maina JN. Morphometrics of the avian lung: The structural-functional correlations in the design of the lungs of birds. Comparative Biochemistry and Physiology. 1993;**105A**:397–410

[56] Maina JN, King AS, Settle G. An allometric study of the pulmonary morphometric parameters in birds, with mammalian comparison. Philosophical Transactions of the Royal Society of London. 1989;**326B**:1–57

[57] Maina JN, King AS. A morphometric study of the lung of the Humboldti penguin *(Spheniscus humboldti)*. Zentralbrat Veterinary Medicine C, Anatomy, Histology and Embryology. 1987;**16**:293–297

[58] Weibel ER. The Pathways for Oxygen: Structure and Function in the Mammalian Respiratory System. Harvard: Harvard University Press (Mass); 1984

[59] Weibel ER. Lung morphometry and models in respiratory physiology. In: Chang HK, Paiva M, editors. New York: Marcel Dekker; 1989. pp. 1–56

[60] Weibel ER. Morphometry: Stereological theory and practical methods. In: Gill J, editor. Models of Lung Disease: Microscopy and Structural Methods. New York: Marcel Dekker; 1990. pp. 199–251

[61] Gehr P, Mwangi DK, Ammann A, Maloiy GMO, Taylor CR, Weibel ER. Design of the mammalian respiratory system: V. Scaling morphometric diffusing capacity to body mass: Wild and domestic animals. Respiration Physiology. 1981;**44**:61–86

[62] Roughton FJW. The average time spent by the blood in the human lung capillary and its relation to the rate of CO_2 uptake and elimination. American Journal of Physiology. 1945;**14**:3621

[63] Roughton FJW, Forster RE. Relative importance of diffusion and chemical reaction rates in determining rate of O_2 exchange of pulmonary membrane and volume of blood in lung capillaries. Journal of Applied Physiology. 1957;**11**:290–302

[64] Staub NG, Bishop JM, Forster RE. Importance of diffusion and chemical reaction rates in oxygen uptake in the lung. Journal of Applied Physiology. 1962;**17**:21–27

[65] Sackner MA, Greeneltch D, Heiman MS, Epstein, LS, Atkins N. Diffusing capacity, membrane diffusing capacity, capillary blood volume, pulmonary tissue and cardiac output measured by a rebreathing technique. American Review of Respiratory Diseases. 1975;**111**:157–165

[66] Weibel ER. Morphometric estimation of pulmonary diffusion capacity. I. Model and method. Respiration Physiology. 1970/71;**11**:54–75

[67] Weibel ER, Federspiel WJ, Fryder-Doffey F, Hisia CCW, Konnig M, Stalder-Navarro V, Vock R. Morphometric model for pulmonary membrane diffusing capacity. Respiration Physiology.1993;**93**:125–149

[68] Cotes JE. Lung Function: Assessment and Application in Medicine. 2nd ed. Oxford: Blackwell Scientific; 1968

[69] Andersen HT. Physiological adaptations in diving vertebrates. Physiological Reviews 1966;**46**:212–243